DATE DUE

The
Reference
Shelf

Space Exploration

Edited by Christopher Mari

The Reference Shelf
Volume 71 • Number 2

The H.W. Wilson Company
New York • Dublin
1999

The Reference Shelf

The books in this series contain reprints of articles, excerpts from books, addresses on current issues, and studies of social trends in the United States and other countries. There are six separately bound numbers in each volume, all of which are usually published in the same calendar year. Numbers one through five are each devoted to a single subject, providing background information and discussion from various points of view and concluding with a subject index and comprehensive bibliography that lists books, pamphlets, and abstracts of additional articles on the subject. The final number of each volume is a collection of recent speeches, and it contains a cumulative speaker index. Books in the series may be purchased individually or on subscription.

Visit H.W. Wilson's Web site: www.hwwilson.com

Library of Congress Cataloging-in-Publication data

Space exploration / edited by Christopher Mari
         p. cm—(Reference shelf; v. 71, no. 2)
    Reprinted articles
    Includes bibliographical references and index
1. Astronautics. 2. Outer Space—exploration   I. Mari, Christopher
II. Series
TL791.S65   1999   99-25424
919.9'04—dc21   CIP

Printed in the United States of America

Contents

Preface

When President John F. Kennedy declared in 1961 that the United States should commit itself to sending a man to the moon by the end of the decade, such an endeavor seemed an extremely daunting task. After all, in its competition with its cold war adversary the Soviet Union, the United States had already lost the race to launch the first artificial satellite into orbit, as well as the contest to put the first man in space. Though Kennedy did not live to see Neil Armstrong mark the surface of the moon with his boots on July 20, 1969, many people around the world felt that the moon landing vindicated the president's belief in human ingenuity. It appeared that human beings were about to embark on our greatest adventure, to leave the comfortable nest of the Earth and journey out towards the stars. The development of a colony on the moon appeared not only inevitable, but quite probable in our near future. Many of the people who had watched the moon landing on television felt the solar system would soon be explored—and very likely colonized to a great extent—as the new millennium approached.

Such predictions, however, proved to be somewhat optimistic, and did not take into consideration political, social, and economic changes. In 1972 President Richard Nixon, declaring that the United States had scored a major victory over the Soviet Union with the moon landings, cut funding for the space program. Politicians, in an ever-tightening fiscal noose, were unwilling to spend more money on the space program, which seemed to have no overt merits. The average American citizen grew more concerned with the social upheaval created at home by civil rights demonstrations and the war in Vietnam than with the long-term potential of space exploration. Economic factors also came into play: the American economy was hit with inflation and recession, which escalated the costs of every mission. With all the belt-tightening, many citizens felt there simply didn't appear to be any pressing need to continue manned programs, especially when those funds could help with so many problems at home. The American space agency NASA (National Aeronautics and Space Administration) committed itself to exploring the other planets in our solar system with cheaper unmanned probes, hoping that someday the political tides would turn and more money would again be granted to send people on journeys to Mars and other celestial bodies.

Yet even with all the political grumbling and escalating costs in an ever-shrinking budget, NASA has achieved some great successes since its glory days in the late 1960s; most notably, the 1981 launch of the space shuttle *Columbia*, the world's first reusable space vehicle. NASA's Soviet counterparts had their share of successes as well, particularly the 1986 launch of the first component of the Mir Space Station, which is still in operation today. Still, many of the great exploratory accomplishments of the last quarter century have come through the aid of robot probes rather than human beings. Since the early 1970s astronauts and cosmonauts have been relegated to low-Earth orbit to perform routine tasks, conduct experiments, and release satellites.

Since the collapse of the Soviet Union in 1991, the once predominant Russian Space Agency has suffered even more than the United States from the economic burden of space travel. Mir, once the crown jewel of the Soviet space program, has since been cursed by the type of malfunctions that come with aging equipment. NASA, still suffering from budgetary restraints as well, has joined forces with the Russians in the hopes of building a new space station. In late 1998 the first component of a truly *international* space station was launched. When completed, 15 countries will have contributed components and manpower to the International Space Station (I.S.S.). It will be as large as two football fields lined up end to end and will be visible to the naked eye from the Earth. While such a station is a phenomenal feat of engineering, it is also criticized in many countries for coming in overpriced and late. (The Russians, with their continual fiscal problems, have delayed the launch of at least one of their components.) In truth, to many American citizens, the International Space Station is somewhat uninspiring and a far cry from the glory days of the *Mercury* and *Apollo* missions, when many of the astronauts were well-known national heroes and their achievements were examples of human ingenuity and courage. In this age of faceless space exploration, the space station has been yet another project of which the public is only dimly aware.

Recently an unusual thing happened. Senator John Glenn of Ohio, the first American to orbit the Earth, was given permission by NASA to return to space to be used as a test subject in some experiments on aging. At 77, Glenn became a national hero once more by returning to space more than a third of a century after his first trip, becoming the oldest human being ever to orbit the Earth. Suddenly the American space program was given a face again and the nation's attention was once again riveted: more than 250,000 people went to the Kennedy Space Center to watch Glenn's launch aboard the space shuttle *Discovery* and millions tuned in via radio, TV, and the Internet.

Glenn's flight was remarkably well timed, especially when considering how a spate of recent discoveries have piqued public interest in space exploration once again. Two years before Glenn's flight, the first suggestion of extraterrestrial life was detected in the form of fossilized microbes in a rock which fell to the Earth from Mars. This rock in many ways has helped accelerate a new era of exploration. Since its discovery, NASA has sent the successful *Pathfinder* probe to the red planet to look for signs of ancient life. A pair of new probes journeyed to Mars in 1999 to further map and study the planet. All of them are part of NASA's program to study the planet over the next decade and possibly send a manned mission to Mars early in the next century. In addition, an ocean has been discovered under the frozen surface of the Jupiter moon Europa, which many scientists have speculated could harbor signs of life, either now or sometime in its history. The orbiting Hubble telescope has detected "wobbles" in the motion of distant stars which indicate the existence of planets far removed from our solar system. The universe seems at once smaller and more vast—the possibility of life existing in other places appears closer to being proven than it has ever been, and yet at the same time that life looks as if it forms in circumstances similar to our own, with the aid of water and heat.

This book's purpose is to give the reader an overall sense of space exploration at the end of the 20th century. Because of their proven accomplishments, I have primarily looked at what NASA and the Russian Space Agency have been working on for the last couple of years, though I have also highlighted new international efforts being made in space exploration, beginning with the International Space Station. There are also many in the private sector who wish to join the space race; because this is a relatively new and important factor in the space age, their efforts are recorded here in a separate section. Also, it should be noted that this book does not pretend to be *the* book on space exploration, nor does it purport to skip to the last chapter of space exploration and reveal what is in store for humanity as we journey into the next millennium. Though this book glances back on the first phase of space exploration, its intention is merely to establish the stage we are in and to encourage the reader to anticipate the next.

The book is divided into five sections, covering the major events that have occurred or are occurring in space exploration at the end of the millennium. The first section, "John Glenn's Return to Space," focuses on Glenn, his mission as a test subject for geriatric study, what he means to people around the world, and what the public reaction has been to his mission. (I have included this section on Glenn to give students of space exploration a sense of this first generation of astronauts: Glenn, in many ways, has become the symbolic beginning and end of the first chapter of space exploration.) As we move into the next section, "Exploration of Mars," the reader is presented with the present group of space explorer—men and women who explore the solar system through ground-based computers and radio-controlled probes. The reader is given an overview of NASA's 10-year mission to study Mars through robotic probes, the men and women behind the successful *Pathfinder* mission, and future planned Mars missions, including the creation of an outpost on the red planet. The third section, "The International Space Station," unravels the convoluted history of the I.S.S. and shows how it evolved from an American response to Mir into an international project. I have also included some of the prevailing arguments for and against this massive project. "Private Enterprise and Space Exploration," this book's fourth section, focuses on recent efforts by commercial industries to mine the moon and other celestial bodies for their natural resources--as well as to promote their potential as vacation spots. "New Technologies and Discoveries," the final section of this book, looks at some things on the horizon in space exploration, both the means to get humanity to the stars and the destinations we may journey to in the coming century. This section discusses new types of space vehicles and new propulsion systems, as well as recent hints of extraterrestrial life on Mars and Europa and the discovery of planets in distant galaxies.

I would like to thank Michael Schulze for the opportunity to edit this book, Cliff Thompson and Hilary Claggett for their advice, and Beth Levy and Sandra Watson for their invaluable help in putting this book together. Special thanks to Brian Solomon and Chris Dieckman for listening to me and making suggestions as I worked this thing out.

Christopher Mari
April 1999

I. John Glenn's Return to Space

Editor's Introduction

On February 20, 1962, John Glenn became the first American to orbit the Earth. His trip took a total of just under five hours, and it made him an American hero. During the cold war, there was only one reason for Glenn to undertake such a mission: to prove that Americans were capable of competing with the Soviet Union in space. Thirty-six years later, on November 5, 1998, John Glenn returned to space on an eight-day mission—not to compete against any foreign power—but to study geriatrics aboard the most sophisticated mode of transportation in the world. The world had changed a great deal between these two flights, but one thing remained the same: John Glenn was still able to reel in an audience. In 1998 American citizens across the country paused during their busy days to listen to or watch his flight, just as they had in 1962. The reasons for Glenn's flight might have changed a great deal, but his second trip proved once and for all that people remain fascinated with those who are willing to risk their lives in the unknown.

To hear officials at NASA speak of Glenn's second mission, one would think they had no idea Glenn's mission would generate such publicity. When they discussed the mission in the press, they stressed the value of the scientific tests they were going to run on Glenn, and the potential benefits these tests could have for all humanity. In truth, there is a lot of science behind Glenn's return to space. On his mission, Glenn was the single test subject used to study the similarities between symptoms of old age and those of weightlessness. In both conditions, researchers have detected a loss of muscle and bone, loss of balance, and disruption of sleep. Scientists speculate that some connection between gravity and the aging process exists, since many astronauts who spend long periods in space appear to age prematurely. Researchers believed they could someday counter some of the ailments of aging and space travel by studying an elderly person in a weightless condition. Enter Glenn. As a member of the *Mercury* Seven, the first class of American astronauts, he became one of the most physically examined men in history. Such thorough medical records made him the ideal candidate for study. After passing the same tests required of all astronauts, the retiring senator began training for his last mission—to help millions of older Americans live better and more productive lives.

The first article in this section, Jeffrey Kluger's "Back to the Future," was published in *Time,* just prior to Glenn's second flight. It is included to give the reader a general overview of John Glenn's career, what he means to scores of Americans who saw his first flight, and what he intended to do on his second. The second article, Danylo Hawaleshka's "A Hero in Orbit," published in the Canadian magazine *Maclean's,* also gives something of an overview of Glenn and his mission, but with more of an international angle. It discusses some of the debate that began cropping up around the mission soon after it was announced. Many argued it was a publicity stunt implemented by NASA and intended to drum up support for the agency's waning budget. Others

1

supported the mission, believing a great deal of knowledge could be gathered by such a study in geriatrics, but also because they felt it gave NASA a much needed shot in the arm. The following three articles are all editorials that support Glenn and honor him as a hero, but debate the merits of his mission. In his *Newsweek* article "Eject Button on Cynicism," Jonathan Alter suggests that Glenn, as a member of the World War II generation, has proven the leadership of his generation once more with his flight: "The giants of history were back for one more skirmish, fighting what Glenn calls the 'couch potatoes.' Their final mission is to prove there are no final missions—no age barriers to serving your country and challenging yourself." In "What Happened to Destiny," an essay written for *Time* and published in the same week, Charles Krauthammer bemoans the Glenn mission, stating "how our horizons have shrunk" with regards to space exploration. He goes on to wonder: "What does it tell us that the only flight we have celebrated for the past two decades is one that looks not forward but back?" Andrew Stephen, in "They Still Want Him Now He's 77," for the *New Statesman*, approves of the mission. His argument is simple: the benefits of this scientific study could be enormous, but he also notes that the mission represents a brilliant publicity campaign by NASA as well.

In an age of cynicism and apathy, John Glenn stands out as a unique public figure of the late 20th century. He seems remarkably untouched by the barbs of an often hostile world and loved by the public and press alike. Appearing oddly pure and honorable in a world considerably more tawdry than the one that embraced his first flight, he is almost a man incongruous with the present—ever the "clean Marine" of the *Mercury* Seven. He has remained a genuine hero to people in the United States and around the world, someone who has made the impossible possible. As one of the early astronauts, he helped to usher in the age of space exploration by expanding human knowledge. It is somewhat fitting then that he has helped to close that first chapter by helping to further improve the human condition.

Back To The Future[1]

John Glenn has a curious tendency to fly machines that try to kill him. He flew them in the Marines; he flew them with the Air Force; he flew them as a civilian. And each time he did, the fact that they were trying to kill him never seemed to trouble him much. One telling incident happened in 1953, during the Korean War. A World War II veteran and a long-time combat aviator, Glenn had been assigned to fly F9F Panther jets in an attack squadron running raids out of Pohang. During one especially hellish run, Glenn encountered an unexpectedly heavy barrage of antiaircraft fire. A cloud of shrapnel ripped one bomb from the undercarriage of his Panther, then another. A second blast punched more than 200 holes in the skin of his plane. Glenn struggled for a few moments to keep his wounded aircraft stable and then realized the effort was futile. Keying open his microphone, he called out levelly to the squadron leader, "I'm going to ease out of here."

The leader, who was too far away to spot the flak coming up from the ground, challenged him. "Why?" he asked. "I don't see anything hot down there."

"Well," Glenn answered, more bemused than flip, "the leader normally doesn't." With that, the 32-year-old flyer peeled off for Pohang.

Last week in Houston, John Glenn, the 77-year-old senior Senator from Ohio, was learning his way around another potentially lethal flying machine. Clad in a blue full-body garment shot through with a webwork of cooling tubes, he stepped into a NASA training room at the Johnson Space Center and glanced at a space-shuttle simulator standing in front of him. A technician then helped him struggle into a heavy orange flight suit. Stuffed into the backpack of the 90-lb. pressure garment was a huge load of survival equipment: a life preserver, an emergency food and water supply, a pair of emergency oxygen bottles, a bouquet of rescue beacons and an array of other gear.

There was no chance that Glenn would need any of this equipment during a training session here on solid ground. But on Oct. 29, when he climbs into a mid-deck seat on the shuttle *Discovery* and prepares to rocket into space for a nine-day mission, he'll face a real, if remote, chance that the craft could spin out before it reaches space and wind up in

If anything could rekindle the magic of the vigorous NASA that was— instead of the flickering NASA that is—it might be the return of the man who first lit the agency's fires.

---

1. Article by Jeffrey Kluger from *Time* 152 p40-6 + Aug 17, 1998. Copyright © 1998 Time Inc. Reprinted with permission. With reporting by Dick Thompson/Houston.

the drink. If it does, the septuagenarian Senator will need all the survival hardware he can get.

By rights, Glenn, who is concluding a 24-year political career and easing into senior statesmanship, ought to be beyond such concerns. By choice, he's not. In less than three months—36 years after he blasted into the sky inside the titanium pod of a Mercury spacecraft—he'll return aboard the relatively lavish space shuttle. Even as Congress's August recess begins and the rest of Washington's lawmakers decamp for their favorite vacation spots, Glenn will be in Houston and Florida for his most intensive month of training since being assigned to the mission.

To hear NASA's detractors tell it, Glenn is manifestly unfit for space travel of any kind. Flying into orbit more than a third of a century after he last made the trip, more than a dozen years after most people his age have begun retiring, and only months after the death of fellow Mercury astronaut Alan Shepard illustrated the frailties of even the most resilient flesh, is, they argue, at best showboating and at worst reckless.

Not so, says NASA. Long ago, the agency noticed a parallel between the changes that happen to a body in space and those wrought by aging on Earth. What better way to study this phenomenon than to send an aged astronaut into orbit? And what better aged astronaut than the one who made the country's first trip?

That's the official story. Perhaps more to the point is that back in the 1960s, NASA was a place for heroes. Every time men rocketed into space, they took a greater risk than on their previous flight, reached for a more audacious and dangerous goal—and almost always succeeded. But after the four extraordinary years between 1968 and 1972, when the U.S. was sending crews to the moon, the agency retreated to the familiar backwaters of near Earth orbit. Aside from a few high notes like the Hubble-telescope repair mission and the horror of the *Challenger* explosion, human space travel became downright dull. And with the first components of the NASA-led International Space Station set to launch within months, things seemed likely to stay that way. For a public that had grown to expect great things from NASA, this was pale stuff indeed. If anything could rekindle the magic of the vigorous NASA that was—instead of the flickering NASA that is—it might be the return of the man who first lit the agency's fires.

NASA will never admit this publicly, of course, and when Glenn goes back to the pad next October, he will go as just another crew member, a lowly payload specialist setting off for a week or so of work. But even NASA administrator Daniel Goldin seems to concede that when he inks the name Glenn onto a flight manifest, he writes more than just a name. "There is," he declared the day he announced Glenn's return to space, "only one John Glenn."

By most accounts, John Kennedy is the key to why Glenn still has the itch to fly in space. When Glenn went aloft on Feb. 20, 1962, the U.S. was taking its first toddling steps on its long march to the moon. Although he was 40, Glenn figured he still had a lot of flying ahead of him. When he returned to Earth, he found otherwise. Like any other astronaut, he periodically approached Bob Gilruth, head of the Mercury program, to inquire about his position in the flight rotation; unlike any other astronaut, he was routinely stonewalled. "Headquarters doesn't want you to go back up," Gilruth would say to him, "at least not yet."

At first, Glenn accepted this with a shrug, but as time went by and more and more of his astronaut brothers were chosen for the Gemini and Apollo programs that followed Mercury, he grew increasingly frustrated. Finally, in 1964, he resigned from NASA. "It was only years later that I read in a book that Kennedy had passed the word that he didn't want me to go back up," Glenn says. "I don't know if he was afraid of the political fallout if I got killed, but by the time I found out, he had been dead for some time, so I never got to discuss it with him."

Glenn spent the next decade working in private industry, most notably (and incongruously) as an executive with the Royal Crown Cola company. In 1974 he parlayed his still glittering name recognition into a seat in the U.S. Senate. Even as a member of Congress, he remained smitten with space travel, but as an aging lawmaker who hadn't been in a flight rotation or ready room in years, he accepted the fact that his professional flying career was over. And it was—at least until three years ago.

In 1995 Glenn, a member of the Senate Special Committee on Aging, was paging through a textbook on space physiology when a thought struck him. Doctors had long since identified more than 50 changes that take place in an astronaut's body during weightlessness, including blood changes, cardiovascular changes, changes in balance control, weakening of the bones, loss of coordination and disruption of sleep cycles. As a lay expert on aging, Glenn recognized that these are precisely the things that happen to people on Earth as they grow older. "I figured we could learn a lot if we sent an older person up, studied what the effects of weightlessness were and tried to learn what turns these body systems on and off," he says. And he had an idea of just who that older person should be.

Approaching the space agency directly with a notion this outrageous was, of course, not the way to go. If 20 years in Washington had taught Glenn anything, it was that bureaucratic balance wheels have to be turned gently. He decided to start by

MILESTONES IN SPACE

1965

MARCH 18 Cosmonaut Aleksei Leonov takes the first space walk, a 10-min. tethered excursion outside *Voskhod 2.*

JUNE 3 Edward White II is the first American to walk in space, floating outside *Gemini 4* for 22 min.

1966

MARCH 16 *Gemini 8* makes the first docking ever with another space vehicle, an unmanned Agena rocket stage. *Gemini 8* later malfunctions, forcing America's first emergency landing.

1967

JAN. 27 Flash fire in the *Apollo 1* command module during a test on the launch pad at Cape Kennedy, Fla., kills astronauts Virgil ("Gus") Grissom, Ed White and Roger Chaffee.

APRIL 24 *Soyuz 1* crashes on re-entry, killing Vladimir Komarov. He is the first astronaut to die during a flight.

1968

APRIL 3 Stanley Kubrick's film *2001: A Space Odyssey,* based on Arthur C. Clarke's short story "The Sentinel," is released.

DEC. 21 Launch of Apollo 8, the first manned mission to orbit the Moon.

MILESTONES IN SPACE

1969

JULY 20 "That's one small step for man, one giant leap for mankind."
Apollo 11 astronauts Neil Armstrong and Edwin ("Buzz") Aldrin walk on the Moon.

1970

APRIL 13 *Apollo 13* Moon mission is aborted when an oxygen tank in the service module ruptures. The crew returns safely to Earth four days later.

1971

APRIL 19 The Soviet Union launches the first space station, an orbiting laboratory named *Salyut 1.*
JUNE 30 *Salyut 1*'s first crew is killed when the spacecraft bringing them back to Earth becomes depressurized during re-entry

1973

MAY 14 *Skylab 1*, the first U.S. space station, is launched into orbit 271 miles above Earth. The first of three crews arrives 11 days later for a 28-day stay.

1975

JULY 17 The *Apollo-Soyuz* rendezvous, highlight of the first international manned space mission

1976

JULY 20 *Viking 1* lands on Mars and transmits the first pictures from the planet's surface. *Viking 2* will arrive

contacting a few NASA physicians and asking them, almost casually, if they had ever looked into the astronaut-geriatric parallel. Why, yes, they had, the doctors said. As a matter of fact, they had published a little pamphlet on the topic. Would Glenn like a copy?

Would he ever! Armed with those few scraps of data, the Senator contacted the National Institute on Aging and suggested that the group might want to hold a conference to investigate the phenomenon further. The NIA agreed, and held two meetings during the following year, compiling a mound of research that strengthened the database considerably.

Finally, in the summer of 1996, Glenn was ready. He approached NASA administrator Goldin and formally pitched his case for returning to space. "I told him there are 34 million Americans over 65, and that's due to triple in the next 50 years," Glenn recalls. "And I told him someone ought to look into this." Goldin, savvy about the wiles of flight-hungry astronauts—even flight-hungry astronauts who haven't flown in 34 years—saw medical merit in the argument and offered Glenn a deal. If the science held up to peer review, he promised, and if Glenn could get past the same physical every other astronaut must pass, NASA would seriously consider his proposal. But, Goldin added, "we've got no open seats just for rides."

It wasn't a decision made lightly. In the months that followed, Goldin wrestled with the matter, agonizing over what he considered his John Glenn problem. At one point, he sought counsel from Tom Miller, Glenn's oldest friend and Marine Corps comrade. "'Can you imagine if something happened [during the mission]?'" Miller recalls Goldin asking. "'My heart says yes, but my brain says no.'"

The scientists and doctors were less ambivalent. By early this year, they had finished their preliminary reviews and concluded that both Glenn's science and his health were sound enough to justify the mission. Shortly afterward, on Jan. 15, Glenn was in his Senate office meeting with a group of constituents from Ohio State University when he got word that Goldin was on the line for him.

Stepping into the bullpen of cubicles just outside his door, Glenn picked up the phone and, after some good-natured stalling and suspense building by Goldin, learned that he would indeed be returning to space and that the announcement would be made the next day. Until then, Glenn was to say nothing. The Senator thanked the administrator, hung up the phone and went back to work without a word to his staff. "He's a military man," says press secretary Jack Sparks. "He knows how to take an order."

When word got out the following morning, the reaction was largely positive, particularly in Congress. Glenn will not be the first lawmaker to fly in space. Senator Jake Garn of Utah and

Representative Bill Nelson of Florida both took shuttle rides in the giddy, all-aboard days before the *Challenger* disaster. In the eyes of many, however, Garn and Nelson were mere junketeers, politicians who wangled a trip into orbit largely for the sake of going up—or, in the case of the famously space-sick Garn, throwing up. Glenn is no mere joyrider. "John has worked hard to prepare for this," says Senator Wendell Ford of Kentucky. "He's not doing it for the publicity. He is doing it to make a contribution."

The response was not quite as enthusiastic at home, where Annie Glenn, the astronaut's wife of 55 years, had to be told the news. Having sweated through her husband's 149 combat missions and one five-hour Mercury mission, Annie had long since become accustomed to Glenn's doing outsize things and incurring outsize risks. In the eighth decade of life, however, she justifiably assumed all that was behind her. "Annie was a little cool to the idea to begin with," Glenn confesses. But in the tradition of a military and NASA wife, she listened to his reasons for wanting to return to space, familiarized herself with his mission and then, as she had done so many times before, proceeded to help him train for it.

That training will be something of a new experience for Glenn, who is used to being the captain of any ship he flies. The flight plan for the October mission lists seven *Discovery* crew members, from Curt Brown, the commander, to Steve Lindsey, the pilot, through three mission specialists and two payload specialists. Glenn's is the last name on the list. No sooner did the crew first meet last January than Glenn made it clear that the chain of command was fine with him. "They wanted to call me Senator, and I said no," he says. "I'm coming down here as John. I'm a payload specialist, and Curt's the flight commander—and whenever they forget that, I correct them."

Even a lowly yeoman like Glenn will have his hands full getting ready to fly aboard his new ship. The first time Glenn flew, he was in a mere demitasse of a spacecraft—one with a single window, 56 toggle switches and barely 36 cu. ft. of habitable space. The joke around NASA in that earlier era was that you didn't so much climb inside a Mercury capsule as put it on.

The shuttle, by contrast, is a veritable flying gymnasium, with 10 windows, more than 850 toggle switches and roughly 332 cu. ft. of space for each of the seven crew members. If astronauts got the 36 cu. ft. the Mercury pilots got, the shuttle could in theory accommodate a crew of 64. And Glenn must learn every inch of this new territory. "We're teaching him how to live and how to sleep and how to clean up, just basic habitability in space," says Brown. "Now we go to space to work. We don't go just to survive."

MILESTONES IN SPACE

1979

MARCH 5 *Voyager 1* makes its closest approach to Jupiter, relaying images of the planet and its moons. *Voyager 2* will follow four months later.

JULY 11 After losing altitude for nearly two years, *Skylab* falls out of the sky and crashes, scattering debris from the southeastern Indian Ocean to western Australia.

SEPTEMBER Publication of *The Right Stuff*, Tom Wolfe's portrait of the early days of America's manned space program.

1981

APRIL 12 Inaugural launch of the U.S. space shuttle *Columbia*, the first reusable manned spacecraft.

AUG. 25 *Voyager 2* flies by Saturn, sending home spectacularly detailed pictures of its rings and moons.

1983

JUNE 18 Sally Ride becomes the first U.S. woman in space.

1984

FEB. 7 Using jet backpacks, astronauts Bruce McCandless and Robert Stewart take the first untethered space walks.

APRIL 10-12 Two astronauts from the shuttle *Challenger* perform the first in-orbit retrieval and repair mission, on the failing Solar Max satellite.

MILESTONES IN SPACE

1986

JAN. 24 *Voyager 2* arrives at Uranus.

JAN. 28 *Challenger* explodes 73 sec. after lift-off, killing all seven shuttle crew members, including teacher Christa McAuliffe.

FEB. 20 Soviet space station Mir launched into Earth orbit.

1987

FEB. 8 Mir becomes the first continuously inhabited space station.

1988

SEPT. 29 First U.S. manned space launch since the 1986 *Challenger* explosion.

1989

MAY 4 For the first time, a spacecraft is launched from the shuttle, as Atlantis astronauts send the radar mapping probe *Magellan* toward Venus

AUG. 25 *Voyager 2* reaches Neptune.

1990

APRIL 25 Astronauts on the shuttle *Discovery* place the Hubble Space Telescope into Earth orbit. Astronomers realize almost immediately that its mirror has the wrong shape.

AUG. 10 *Magellan* begins to orbit Venus and relay radar images of its surface as well as other data.

More nerve-racking than mastering what goes on inside the shuttle, though, is mastering what could go on outside. One of the most hair-raising parts of Glenn's training involves emergency escape procedures. Crew members on shuttles must be prepared to ride slide-wire baskets down from the launch pad if a fully fueled shuttle threatens to blow; shimmy down an escape pole and parachute away from the ship in the event of a postlaunch emergency below an altitude of 20,000 ft.; and rappel down ropes from the hatch if the spacecraft makes an emergency landing on tarmac. On his Mercury flight, Glenn's only safety measure was an escape rocket designed to ignite and carry the spacecraft out of danger if his Atlas rocket appeared likely to explode.

Not everything about the shuttle will be more difficult. During the Mercury days, the astronauts pulled a gravity load of up to 7.9 Gs during their ascent, meaning that a pilot like Glenn who weighed 168 lbs. would briefly feel as if he weighed a whopping 1,327. Shuttle astronauts generally pull no more than 3 Gs, and Glenn, who has not added much weight to his still fit frame in the past 36 years, should tolerate that burden easily.

Then too, there are creature comforts aboard the shuttle that the Mercury pilots didn't dare dream about. Glenn's only meal on his first, brief mission in space was a tube of applesauce he sucked through a straw. The shuttle offers a decidedly better bill of fare, including such delicacies as smoked turkey, Kona coffee and dried apricots. All the meals are sealed in plastic packets, each of which is coded with a colored dot to indicate which crew member it is intended for. The color code for Brown, the commander, is red; for Glenn, a payload specialist, it's purple. "The shrimp cocktail they fix is very, very good," says Glenn, "as good as what you'd get at Delmonico's. Curt likes shrimp, and I always tell him that when he's on the flight deck and I'm hungry, I'm going to go looking for a red dot."

But Glenn is going aloft to do more than tuck into the cuisine. *Discovery* will ferry a number of payloads in its cargo bay, including a Spartan satellite that will be released into space to take readings of the sun, a pallet of sensors to measure the ultraviolet environment of space, and several new components for the Hubble Space Telescope that need to be tested in the extreme conditions of space. Most important, the ship is carrying the Spacehab science module, a pressurized laboratory that is connected to the crew compartment and provides additional space for conducting medical experiments. It is here Glenn will be doing most of his work, processing blood and urine samples from the rest of the crew and sitting still for the battery of tests that will be run on him.

Those tests would try the patience of any patient. Throughout the flight, Glenn's heart rate, respiration, blood volume and pressure will be monitored regularly. Doctors on Earth want to analyze his blood for immune function and protein levels, and this will require taking so many samples that throughout the flight, Glenn will wear a catheter implanted in his arm, allowing easy access to a vein without a new needle stick each time. He will wear a suit wired with sensors to measure his sleep cycles and will swallow a horse-pill-size thermometer that will take temperature readings as it passes through his body.

These and other findings will be compared with base-line readings taken before lift-off, which are already being assembled. Glenn routinely walks around the grounds of NASA's Houston facility with monitors strapped to his wrist and belt. When he returns from space, he will face yet another battery of tests, including an MRI to look for changes in his spinal cord and bone-density tests to look for mineral loss. "All of this," Glenn says, "gives us the potential not only of dealing with the frailties of our already aged population but of helping younger people avoid problems as they get old."

Or so NASA says. Not everyone in the space community agrees. Alex Roland, a former NASA historian and chairman of the Duke University history department, has been outspokenly skeptical of Glenn's mission, questioning its scientific value and dismissing it as a trivial or even foolish use of NASA's scarce resources. If critics like Roland are right, the mission's science is merely a fig leaf. If it's a fig leaf, what is it covering? "This space flight is the same as the first one," says John Pike, director of space policy for the Federation of American Scientists. "It had everything to do with making the country feel good. It's about the right stuff, not science. Which is fine with me." Newsman Walter Cronkite, whose coverage of the Mercury missions made him as much of a television icon as the astronauts, agrees that Glenn's upcoming flight "is bringing back a public interest in space flight."

Whether or not this is true, there is no denying that Glenn's 1998 mission will be rich with echoes from his 1962 mission. Once again there will be the program-pre-empting coverage; once again Annie Glenn and her family will be seen watching anxiously as the rocket that carries the head of the household explodes off the ground and falls back to Earth; once again there should be the triumphal return.

The first time Glenn flew, the family stayed at home in Arlington, Va., watching the launch on TV, since the Glenns were reluctant to pull their son and daughter out of school for the trip to Cape Canaveral. This time wife, children and the Glenns' two grandsons will all be there for lift-off. Glenn takes a small, whimsical pleasure in pointing out that his grandsons, who will be 16 and 14 in the fall, are the same age his son and daughter—now 52 and 50—were the last time he flew.

MILESTONES IN SPACE

1993
DEC. 4-10 Astronauts capture the Hubble Space Telescope and repair its optics. To everyone's surprise, the mission is a complete success.

1994
FEB. 3 Sergei Krikalev becomes the first Russian to fly on a U.S. spacecraft.

1995
MARCH 22 Cosmonaut Valery Polyakov returns to Earth after spending a record 437 days 18 hrs. in space aboard Mir.
JUNE 29 As part of America's 100th manned space mission, the shuttle *Atlantis* docks with Russia's Mir.
DEC. 7 After a six-year journey, the *Galileo* probe reaches Jupiter.

1998
OCT. 29 John Glenn, now a 77-year-old Senator from Ohio, is scheduled to go into space a second time, aboard the shuttle *Discovery*.
NOV. 20 First piece of the International Space Station due to be launched.
DEC. 10 Planned launch of the *Mars Climate Orbiter*, which will arrive at its destination in September 1999. It will be followed by the *Mars Polar Lander*, scheduled to be launched in January 1999.

For anyone contemplating Glenn's return to space, this kind of existential ciphering is irresistible. The country is now further in time from Glenn's first trip into orbit, for example, than Glenn's first trip into orbit was from Lindbergh's flight across the Atlantic. A man who was Glenn's current age when Glenn was born would himself have been 17 years old when the Civil War began. Then too, there are the people who saw Glenn's first flight who either will or won't be here for the second. Khrushchev, Kennedy, Johnson, Mao Zedong—all towering figures in 1962, all dust now. Castro—communism's beachhead in the West then, old and isolated now. Queen Elizabeth—young and remote monarch then, old and remote monarch now.

That kind of perspective shifting, that kind of standing back from the pointillist portrait of history, may be what Glenn's return to space is really all about. Glenn and NASA will never wholly concede this spiritual point, but Glenn and NASA don't have to concede it. John Glenn flew in 1962, and an exuberant country decided it just might live forever. Thirty-six years later, an older, more sober nation could use a little of that feeling again.

A Hero in Orbit[2]

Exploration of one kind or another has a long history in and around the rolling hills of New Concord, Ohio, boyhood home of John Herschel Glenn Jr. In the early 1800s, frontiersmen driving eight-tonne Conestoga wagons, pulled by teams of six powerful horses, hauled freight through New Concord on their way to opening up the American West. In 1828, Presbyterian immigrants of Scottish and Irish descent settled the eastern Ohio town bisected by the National Road—for decades, the most heavily travelled artery in the United States. The highway's popularity ebbed with the coming of trains, then surged again with the primacy of the automobile. Finally, in the early 1960s, the new Interstate 70 left Ohio's storied east-west road principally to local traffic.

Still, there was one more frontiersman to come. As one of the seven famed Mercury astronauts—celebrated in the book and movie *The Right Stuff*—Glenn rocketed into history in 1962 inside the tiny *Friendship 7* capsule, becoming the first American to orbit the Earth, and a certified Yankee hero. But now, on the eve of his second voyage into space—at the extraordinary age of 77—his reception has been decidedly mixed. Critics say his inclusion in the seven-person crew aboard the space shuttle *Discovery*, scheduled to lift off this week from the Kennedy Space Center in Florida, amounts to no more than a publicity stunt by NASA, the beleaguered, perennially cash-strapped U.S. space agency. They question the scientific value of letting an aged hero hitch a nine-day joyride on a shuttle mission that could cost as much as $600 million.

But NASA says this is no joyride. In the agency's view, monitoring the oldest man ever to fly in space could increase its understanding of microgravity's adverse effects on astronauts, effects that physiologically mimic aging's assault on earthly bodies. Such research, NASA officials add, could also translate into better health care for the elderly back on Earth. It is a position seconded by many in Glenn's postcard-pretty home town, where staunch supporters like Lorle Porter, a professor emeritus at New Concord's Muskingum College, say there is nothing wrong with an active septuagenarian going into orbit. "It's not," the historian says, "like he's some old geezer going up in space to stare out the window."

As far as NASA is concerned, Glenn could scarcely have come along at a better time. After 91 shuttle missions since

In [NASA's] view, monitoring the oldest man ever to fly in space could increase its understanding of microgravity's adverse effects on astronauts, effects that physiologically mimic aging's assault on earthly bodies.

2. Article by Danylo Hawaleshka. From *Maclean's* p74-6 Nov. 2, 1998. Copyright © 1998 *Maclean's*. Reprinted with permission.

While typical shuttle flights draw about 150 requests for accreditation at the launch site, Glenn's mission, STS-95, has attracted more than 4,000 from around the world.

1981, space travel has become so routine that launches barely merit a blip on the evening news. But Glenn, who has spent the past 24 years as a Democratic senator, changed all that by successfully lobbying NASA for another trip. The mutually beneficial result has Glenn getting what he wants, and NASA and partners like the Canadian Space Agency benefiting from the public's renewed interest in the space exploration saga.

That interest and, more importantly, the financial support that space program advocates hope will flow from it, will be crucial in the coming years as NASA and its partners prepare to assemble the oft-delayed International Space Station. The station's first Russian-built module is to be launched from the Baikonur Cosmodrome in Kazakhstan in less than a month. And Canadian astronauts will play a crucial role in its assembly, with three of them—Julie Payette, Marc Garneau and Chris Hadfield—scheduled for separate shuttle missions beginning next year. With the world now watching, the stage is set for an unprecedented international exploration of the vast frontier that is space.

Science aside, picking the senator was, in many ways, an inspired decision. In New Concord, (population 1,200, not including the college students), Glenn is widely regarded as a genuinely nice, down-to-earth person. He believes in God, has integrity, a profound respect for his country and a Presbyterian commitment to public service. In the 1960s, he showed the right stuff in a number of ways, including scolding his hotshot Mercury astronaut colleagues for chasing fast women and running the risk of embarrassing NASA. Perfect American-icon material—then and now.

And the media have taken the bait. While typical shuttle flights draw about 150 requests for accreditation at the launch site, Glenn's mission, STS-95, has attracted more than 4,000 from around the world. Since NASA announced in January that Glenn was physically fit to fly, the popular politician has been on the covers of *Time*, *Life*, *Newsweek* and *Popular Science*. CNN even hired television legend Walter Cronkite—the voice of space exploration in the 1960s—to co-host its mission coverage, including the live broadcast of the launch.

The attention is well deserved, Ellis Duitch would say. Now a spry 95, Duitch was Glenn's high-school science teacher in New Concord. He knows that inquiring reporters are eager to find a chink in Glenn's shining suit of Americana. He volunteers "The Skunk Story," even though he knows it will not quench the media's thirst for scandal. As Duitch tells it, Glenn's future wife, Annie Castor, and her teenage girlfriends were attending an evening function at Brown Chapel on the grounds of Muskingum College. Glenn, all red hair and freckles, was accompanied by one of his buddies. They were eager to walk the girls home. The boys stumbled upon a skunk in the church basement and poked it with a stick until it sprayed. "He cut loose, cut good spray," Dutch says with a laugh. "It wasn't very long until that meeting broke up, but I don't know if they ever got to walk any of the girls home." And that's as racy as it gets.

For all Glenn's wholesome reputation, NASA knows that even the nicest guy in the world, or space, has to earn his keep. Glenn will be one of two payload specialists, essentially scientists in space. Glenn's primary subject: himself. A battery of tests he is to conduct on his body will investigate the similarities between aging's impact and what astronauts go through when subjected to the near-absence of gravity. Phenomena under examination include bone and muscle loss, balance disorders, sleep disruptions and a weakened immune system. Glenn will be required to, among other things, monitor his heartbeat and collect samples of his own blood and urine. NASA scientists, meanwhile, will record his brain-wave activity while he sleeps, and his co-ordination while he is awake.

There is, however, an inescapable problem: the value of studying just one person has profound statistical limitations in terms of the conclusions that researchers can reach. Canadian astronaut Dr. Dave Williams, a member of a shuttle mission last April and now director of the Space and Life Sciences Directorate at NASA's Johnson Space Center in Houston, readily acknowledges the limitations of what Glenn and NASA can accomplish. Nevertheless, while briefing reporters in Toronto in September, Williams maintained that Glenn's participation is an important first step that will "open the window of understanding to a new world."

Time will tell. In the interim, there will be plenty for Glenn's six crewmates to do. Using the mechanical Canadarm, they will deploy a satellite, the Spartan 201, to observe the sun's outer atmosphere in an effort to understand more about the solar winds that can damage communication and television satellites. They will also test new components to upgrade the Hubble Space Telescope and study ultraviolet radiation.

Research on the mission has a distinct Canadian component. John Davies, a professor with the University of Toronto's Institute of Biomaterials and Biomedical Engineering, has designed an experiment to study the primary cells responsible for bone formation and bone loss. What people often do not realize, Davies says, is that the human skeleton is a dynamic tissue, with bone constantly being made and eaten away. Over the course of about 15 years, the average person's skeleton will be completely replaced with new bone. In space, the balance is lost between bone-making cells and cells that break down bone. Davies wants to find out whether the imbalance is caused by underactive bone-making cells, overactive bone-eating cells, or a third and as yet unknown culprit. There is a pressing need to find out. After just three months aboard space station Mir, cosmonauts lose almost 20 percent of bone mass around hip joints. "Imagine," says Davies, "what would happen in a two and a half-year trip to Mars."

What people often do not realize, Davies says, is that the human skeleton is a dynamic tissue, with bone constantly being made and eaten away.

A second Canadian study, designed by Louis Delbaere, head of biochemistry at the University of Saskatchewan, will attempt to grow protein crystals of an enzyme that makes too much glucose in diabetics. In the near-weightlessness of space, crystals grow bigger and more precisely than they do on Earth. When returned from space, the crystals will be X-rayed by researchers to determine their three-dimensional structure, with the aim of developing a drug to block the enzyme. In another study, pathologist Don Brooks of the University of British Columbia in Vancouver has designed an experiment to separate cancer cells from normal cells, a process that could lead to medical advances on Earth. Operating in the microgravity of space, Brooks says, will give his research a large boost. "It's an efficient way," he says, "of answering all sorts of questions."

Questions of another kind surround construction of the fabulous International Space Station, a permanently manned platform for space experimentation. First and foremost is how much will it cost. The United States conservatively estimates assembly costs for the station at $27 billion (plus $19 billion to operate for the first decade after it is completed). But major mission failures, where an entire payload is lost, or other unforeseen cost overruns, could push the figure several billion dollars higher. As it stands, NASA estimates it will take 45 missions by its space shuttles and Russian craft to assemble the station by 2004.

Another question is timing. Last month, NASA confirmed that the station's first element—the Russian-made Zarya Functional Cargo Block module—will be launched on Nov. 20, signalling the start of the station's long-awaited construction. Zarya had been scheduled to fly last June, but the cash-strapped Russians were unable to meet their deadline. NASA also confirmed that the second element, the U.S.-made Unity module, is to be launched on Dec. 3.

It is the third component, Russia's so-called service module, that is currently causing anxiety. That module will house the station's first crew, scheduled to be led by an American accompanied by two cosmonauts. Its launch had been planned for next April, but delays have forced it to be rescheduled for July, putting off plans to have the station inhabited next summer. The service module's three-month postponement will almost certainly push back the shuttle plans of Canadian astronauts Payette, who was expecting to fly in May, Garneau who was set to go up in August, and Hadfield, who was booked for December, 1999. All three missions include components related to the space station's assembly. "It's safe to say that they'll be delayed," says NASA spokesman James Hartsfield.

While there have been delays—and more appear inevitable given Russia's economic and political instability—the station's assembly seems destined to proceed. A critical tool in the construction will be the next generation of the robotic Canadarm that has performed so well on shuttle missions. Built by Toronto-based Spar Aerospace, the so-called Space Station Remote Manipulator System will be equipped with a Special Purpose Dexterous Manipulator. In other words, a new and improved Canadarm, featuring a "Canada hand" to move equipment and supplies, release and retrieve satellites and aid astronauts with assembling the station. The first element of the new system was to have flown with Hadfield at the end of 1999, but that flight will likely be postponed until early 2000.

But that is all in the future; this moment belongs to Glenn. Whether his mission yields any significant science or not, there is no denying the resonance of his return to space. Harold Kaser, a retired Presbyterian minister who played college football with Glenn at Muskingum, wishes the home-town boy well, recalling the sentiments of Glenn's back-up pilot. "Using the words that astronaut Scott Carpenter used: 'Godspeed, John Glenn.' Do it again," Kaser says. Judging by the worldwide attention to Glenn's mission, this is one American hero who travels well.

Whether his mission yields any significant science or not, there is no denying the resonance of [Glenn's] return to space.

Eject Button on Cynicism[3]

Some firsthand experiences fail to live up to expectations: A shuttle launch is not among them. In the press grandstand where I watched *Discovery* rise against the cloudless sky, the media hit the abort button on cynicism. The Earth shook to the sounds of man, three miles away. The candle lit. When wide eyes narrowed again, hard questions about the future returned. But only someone stripped of awe can leave a launch untouched.

And yet the exhilaration was tinged with bittersweet yearning. This was, after all, a sequel, not vivid history itself: low orbit, not Mars. John Glenn saw his dreams made real again—would the rest of us? Or was this just the networks giving us a break from tawdry (and increasingly low-rated) Washington news, a Mark McGwire moment? A hero sandwich for a hungry nation?

John Glenn saw his dreams made real again— would the rest of us?

No, it was more than manufactured. The nostalgia conjured a real sense of national purpose and performance. In the NBC News booth, I met Scott Carpenter, whose 1962 benediction, "Godspeed, John Glenn," still tingles every time. "I guess I said that because no one had ever gone that fast before," he explained. Repeating his send-off on the air, he had tears on his face, once the wrong stuff for stoic astronauts, now right for the occasion. Baseball legend Ted Williams, confined to a wheelchair, made the trip to the Cape; he and Glenn had flown combat missions together. The day felt a bit like it must have at Fenway Park in 1960, when Williams hit a home run in the last at-bat of his career.

President Clinton and a good chunk of Congress came down; it's the new hip junket, assuring NASA funding in a recession. And they knew the press was here. Saturation City. We communicate so much more nowadays about so much less. I met an old space reporter from the New York *World-Telegram* named Richard Slawsky. When Alan Shepard became the first American in space in 1961, Slawsky said, there were 39 telephones for the press corps at Cape Canaveral. This week there were 5,000 accredited reporters, most with cell phones, plus hundreds more lines.

The classic Greek hero is young and brave—like Glenn the first time—but older, safer heroism has some practical advantages. If the odds of glory are smaller, so are those of disillusionment. We know John Glenn is not going to get

3. Article by Jonathan Alter. From *Newsweek* p28 Nov. 9, 1998.
Copyright © 1998 Newsweek, Inc. All rights reserved. Reprinted with permission.

arrested for date rape; the pride won't be betrayed. If the scientific justification for his trip is a little fishy, no one much seems to care. Besides, we wouldn't want someone asking what *we* bring to the party when the chance finally comes to hitch a ride.

And that was the inspirational corner turned last week—if a 77-year-old could do it, so could we. In that sense, Glenn now represents a healthy ratcheting down of the heroism required for space travel. This flight is helping to exorcise the ghosts of the *Challenger* disaster, which had reintroduced fear into the mental baggage Americans carry into space. After all, by the standards of 20th-century aviation, the normalization of space has been agonizingly slow. Glenn's first spaceflight, 36 years ago, was in turn nearly 36 years after Charles Lindbergh first flew solo across the Atlantic. Imagine if an elderly Lindbergh had, in 1962, offered to reprise his famous flight in a bigger, fancier plane. What a yawn. By then, passenger-jet travel across the Atlantic was already routine.

Now, with the coming commercialization of the space program, limited tourism in space may actually be less than a decade away. Currently, it costs NASA roughly $10,000 for every pound of payload. That makes the market price for launching a human being something around $2 million, which more than a few wealthy Americans would gladly spend for the privilege. That doesn't include capital costs, and there aren't any new shuttles on the way. But with the creation in 1996 of the United Space Alliance (USA), a joint venture between Boeing and Lockheed Martin to be NASA's sole contractor, the payload price should start falling. If all goes according to plan, NASA will soon become a supervisory and research-and-development agency, no longer in direct operational control.

But first comes the space station, which will begin going up aboard the shuttle *Endeavor* in December, a far more consequential mission than this one. I toured a reproduction of it at Cape Canaveral and heard the descriptions of growing crystal proteins and converting astronaut urine into drinking water. Not surprisingly, the only question the officials wouldn't answer was the price tag, which is now estimated at close to $100 billion. Needless to say, the cheap engineering that worked so well for the Mars *Pathfinder* has not been applied to the space station, which is apparently being built in the priciest way possible. The most controversial portion of the project's budget is the money to prop up the Russian space program. Actually, that's the spending I find the most defensible, since out-of-work Russian scientists scare me.

This flight is helping to exorcise the ghosts of the Challenger *disaster, which had reintroduced fear into the mental baggage Americans carry into space.*

In 220 years, this country has produced only two monumental generations of leadership—the Founding Fathers in the late 18th century and the World War II veterans in the mid-20th. The movies evoked at Cape Canaveral last week were not just *The Right Stuff* and *Apollo 13* but *Saving Private Ryan*. The giants of history were back for one more skirmish, fighting what Glenn calls the "couch potatoes." Their final mission is to prove there are no final missions—no age barriers to serving your country and challenging yourself. That means not just more septuagenarian space travelers but more tutors and child-care workers and counselors, perhaps tens of millions more local heroes as the baby boom retires, in a vast autumn harvest of experience.

What Happened to Destiny?[4]

I hate to be the skunk at the picnic—or rather, the great national celebration that attended the return of John Glenn to space. But amid all the high-fiving about how wonderful and glorious it was, we seem to have glossed over the fact that on that beautiful Thursday morning we sent the same man on the same trip he made 36 years ago. It is as if we had a great big back-slapping national jamboree at Kitty Hawk in 1939 to watch the Wright brothers skim the sand in a new biplane.

Don't get me wrong. This is no knock on John Glenn. John Glenn is a hero. He deserves to go anywhere he wants to go. He would have earned his wings if he'd done nothing more than fly those 149 combat missions in World War II and Korea, let alone risk his life as the first American to orbit the earth.

This is no knock on NASA either. Sure, it shamelessly hyped the Glenn flight. But NASA, a living rebuke to those conservatives who believe that government can do nothing right, is perfectly entitled to use whatever stunt it can to gin up the nation's constantly flagging interest in the greatest human achievement of this half-century, space flight.

And what a job it did with Glenn. When was the last time the country stopped, riveted to the TV, anxiously awaiting a launch? When was the last time television sets were wheeled into classrooms? When was the last time kids cheered our national space truck making yet another haul? Most people were surprised to learn that the Glenn flight is the shuttle's 92nd. Except, sadly, for *Challenger*, who can remember any of the other 91?

No, the knock is on us, the nation that John Kennedy once galvanized with the challenge "of landing a man on the Moon and returning him safely to Earth," impossibly, within 8 1/2 years. Once, we had the spirit to go. And we did. Then we came back. And ever since, we've sat.

Maybe the critics are right. Maybe what animated us back then was less the spirit of exploration than the spur of nationalism. Maybe it was all about beating the Russians. How else to explain how we've been content to go around in circles—literally, around and around in low-earth orbit—for the past quarter-century?

Let's be honest about the Glenn trip. The attempt to sell it as science, though entirely understandable, is entirely laugh-

> *When was the last time the country stopped, riveted to the TV, anxiously awaiting a launch? When was the last time television sets were wheeled into classrooms?*

---

4. Article by Charles Krauthammer. From *Time* p130 Nov. 9, 1998.
Copyright © 1998 Time, Inc. Reprinted with permission.

able. This enormous expense—and considerable risk—to pick up a datum or two about geriatrics? How our horizons have shrunk. Space flight was once about destiny, not telemetry. Three decades ago, Kennedy spoke for the nation when he ringingly declared, "We choose to go to the moon." What have we to say now? "We choose to study Metamucil digestion in microgravity?"

It is not as if we have nowhere to go but endlessly around the Earth. True, until a few years ago, it could have been argued that a Moon base was impractical and Mars exploration even more so. But we have recently discovered ice on the Moon, which makes the provision of water and (from water) fuel a real possibility. Similarly, new ideas have been advanced for using Mars' water and pre-positioned fuel stores to reduce radically the loads human travelers would have to haul there.

In the early 1970s, we had a window to the Moon. We let it close. Amid the retreat and demoralization of the Vietnam era, we came home, America. The window can be reopened, but it requires an act of will.

In the early 1970s, we had a window to the Moon. We let it close. Amid the retreat and demoralization of the Vietnam era, we came home, America. The window can be reopened, but it requires an act of will. Instead, we are now set to spend the next decade or two building a cozy little house, called a space station, in near-Earth orbit.

I know it sounds quaint, but what happened to destiny? What happened to the lust for the frontier? Perhaps in any age of exploration a stage of natural exhaustion sets in after the first burst of discovery. After all, Columbus sailed in 1492. Yet it was not until 73 years later that the first permanent European settlement was established in North America. What a pity if we had to wait 73 years before establishing a living human presence on the moon.

I love John Glenn. And his flight may yet advance exploration—not because of the science he brings back but because of the enthusiasm he generates. It is no fault of his that the enthusiasm is of a peculiar kind. Space flight is designed to evoke the edgy, restless feeling of a beckoning future; this flight evokes feelings of warmth and nostalgia, "that moment when we could feel young again," as Tom Shales of the *Washington Post* put it. What does it tell us that the only flight we have celebrated for the past two decades is one that looks not forward but back?

Glenn's flight is testimony not to the hold that space has on us but to the esteem in which we hold Glenn. It is a salute not to daring but to celebrity. Had some obscure 77-year-old astronaut trainee gone up instead, there would have been nothing approaching the current hoopla. The story of this flight is that a man with a name—a hero—went into space. The point of space exploration, however, is for a man with no name to do something so magnificent, so improbable, so epochal, as to enter the pantheon of heroes. As happened 36 years ago to a man called John Glenn.

They Still Want Him Now He's 77[5]

I first met John Glenn about seven years ago, and my immediate impression was unchivalrous: he's an old man!

He was then just 70, but seemed older and more doddery than most people his age. Whether I'd subconsciously been expecting a super-fit, crew-cut superhero—the 40-year-old John Glenn who became the first American to orbit the Earth, on 20 February 1962—I don't know; even his two children, I had to remind myself, were older than me. But next Thursday the 77-year-old trooper will blast off again from Cape Canaveral in one of the most intriguing space adventures NASA has ever pulled off.

A week or two ago I happened to be in NASA HQ, and one of the top people there—a former astronaut who has been up three times, even going for a space walk—told me privately that, contrary to widespread media assumption, the Glenn mission is viewed by the tightly knit space world with enthusiasm rather than disapproval: "I had 155 hours up there," my man told me. "John had only four." (Four hours, 55 minutes and 23 seconds, to be precise.)

By becoming an instant American hero, in fact, John Glenn's career suffered calamitously: President Kennedy immediately classified him as too valuable a commodity to be risked again and so he missed the opportunity ever to walk in space or go to the Moon.

In the old days, Glenn was sent aloft by one of Wernher von Braun's 93-ft Atlas rockets in his *Friendship 7* capsule, which now sits for us all to see a mile or so away from where I live in Washington: it looks tiny (9ft by 6ft) and flimsy and it lacked even the most basic computerised aid that we take for granted today.

But Glenn nonetheless encircled the globe three times in it, at a speed of 18,000 miles per hour and an altitude of 160 miles.

Next week a rocket double the length but with 20 times the thrust will fire him twice as far from Earth, and this time he will make 144 orbits; there will not only be 800 separate control switches inside the space shuttle, but six other astronauts (including, God forbid, a woman) and even a drawerful of junk food should the old chap become hungry.

---

5. Article by Andrew Stephen from the *New Statesman* p30 Oct. 23, 1998. Copyright © 1998 the *New Statesman*. Reprinted with permission.

Though the cartoonists have had a field day—my favourite shows Glenn going for a space walk on a Zimmer frame— there is (despite all the jeers to the contrary) a serious side to this mission. Even though 379 people (239 of them Americans) have now been in space—and next week's shuttle flight will be its 92nd—there has been remarkably little research into the reaction of the human body to the weightlessness of space.

It has long been observed that, even after only a few days in space, a human body will lose as much as 10 percent of muscle tissue—with similar drops in the calcium and other mineral levels of bones. It's the same with animals, too: NASA doesn't like to publicise this kind of fact, but in one of the four shuttle missions flown this year, half the animals taken aboard came back dead.

What has not been seriously researched is why this should all be so. Scientifically, the intriguing coincidence is that in space even a fit and relatively youthful body deteriorates in much the same way as it does on earth.

That is one of the reasons why nobody has yet seriously contemplated sending a human to Mars, because in a return journey taking years rather than weeks the condition of the astronauts' bodies would deteriorate too quickly. In short, they would age prematurely.

You read a lot about how super-fit John Glenn still is; but could this be a reason, I can't help wondering, why he looked old to me close-up when he was only 70?

He is, actually, the most medically documented astronaut in history, having first been scrutinised when he joined the fledgling Mercury team in 1959. This year he has been subjected to intense medical examination again, and will spend most of his nine days in space with 21 electrodes stuck to his chest.

He will swallow a probe that will send back data about what space is doing to his insides. The female colleague, a Japanese cardiologist, will take at least one blood sample a day. Urine and feces will be retained for laboratory analysis (one of Glenn's routine tasks is to attend to this chore for the entire crew). For the $477 million that every shuttle mission costs, in fact, NASA hopes it will achieve more than a public relations coup with the Glenn mission.

Data on Glenn will be just that: data on one particular man. But, despite what you will be reading in the coming days, next Thursday's space shot is not just a gimmick to stir up enthusiasm for NASA's flagging budget—nor is it a reward for a decent but uninspiring (and about-to-retire) Democratic Senator for 24 years (the joke about him is that

Even though 379 people . . . have now been in space . . . there has been remarkably little research into the reaction of the human body to the weightlessness of space.

he always gets more applause when he stands up to speak than when he sits down afterwards).

The fact is that when Glenn went up in 1962, there were 17 million Americans aged 65 or over; the number today is double that. In only three decades' time, oldies will exceed 69 million. If a 77-year-old John Glenn can succeed in shedding any light on why the bodies of astronauts and old people begin to fail in the same way—he being the first to qualify on both counts—he could, notwithstanding the cynical naysayers, become even more of a 20th-century American hero than he is already.

I look forward to seeing him jogging near his home in Bethesda—well, let's call it fast walking—and sailing that 66-ft boat of his in Annapolis when he returns next month.

In those melodramatic but none the less sincere words of NASA control when he shot off into the unknown 36 years ago: God speed, John Glenn.

II. Exploration of Mars

Editor's Introduction

Mars, that bright red object in our night skies, has long been an object of fascination for human beings. For untold millennia the human race has looked up towards that heavenly body and tried to understand its secrets. Named after the Roman god of war because of its blood-red color, Mars was believed to be a great star in the sky whose risings and settings portended the future. After some time—and with the development of telescopes—people realized it was neither a god nor a star, but a planet. Yet human fascination with the red planet only grew stronger, so we devised new ways to reveal its mysteries.

Beginning in 1877, an Italian astronomer by the name of Giovanni Schiaparelli started studying Mars in great detail with his telescope. He soon observed a vast network of what he called *canali* crisscrossing the planet's surface. This Italian word can be used to describe either natural waterways or artificial ones. Percival Lowell, a Boston millionaire who was fascinated with astronomy, believed them to be the latter. In 1894 he set up an observatory on Mars Hill in Arizona, bringing with him by train a 24-inch refracting telescope. For many years Lowell looked through his telescope and sketched out what he thought he saw on the Martian surface. According to his theory, Mars was a dying planet, drying slowly, and its indigenous intelligent life constructed these canals in order to bring life-saving water from their polar ice caps.

Though Lowell was wrong about the canals, he has recently been proven correct in observing that Mars was once a more watery planet. Yet Lowell's greatest contribution to Mars research was not so much his predictions, but the fascination he inspired in other people. In 1898, the English novelist and philosopher H. G. Wells published *War of the Worlds*, a fantastical science fiction novel that described a Martian invasion of the Earth. In 1938 Orson Welles adapted it into a radio play and caused a panic across America when people began to believe they were listening to an actual Martian invasion. More science fiction novels and films followed and we became more curious about visiting our next door neighbor.

When the space age dawned in the middle of the 20th century, American scientists at NASA didn't wait long to send an early deep space probe to Mars. In 1964, *Mariner 4* flew by Mars, sending back the first photographs and scientific information about the red planet. In 1969, *Mariner 6* and *7*, two more fly-by missions, followed. In 1971 *Mariner 9* became the first orbiter to study and map the planet in great detail. The information received from each probe confirmed what earlier studies had postulated: Mars was a dead dry planet with a carbon dioxide atmosphere—quite inhospitable to life as we know it. The surface features, however, appeared to have been shaped by water at some point in the planet's history. And water suggested the possibility of ancient life. With that in mind, NASA sent the *Viking 1* and *2* probes in 1976. Each was equipped with an orbiter and a lander to study the planet's surface and search for hints of extraterrestrial life in its topsoil. Both landers were programmed to scoop up and test soil samples for organic material, but neither recovered any evidence of past life.

With the exception of the *Phobos* orbiter, sent by the Soviet Union in 1989, no artificial probe was sent to Mars in nearly 20 years. Recently the United States has embarked on a new and more ambitious study of Mars, designed to study the planet by robotic probes over the next decade. This second section is devoted to these new probes and the revolutionary ways in which NASA is conducting the missions. In 1993 the expensive and ambitious *Mars Observer* was launched, loaded with equipment to study the planet. After taking a single black-and-white eight-by-ten glossy, the probe went silent. The *Observer* ended a long period of NASA's history—designing probes over a decade and loading them with a slew of instruments. Instead, the space agency is now looking to do things "better, faster, cheaper," in the words of NASA chief Dan Goldin. In essence, the new probes are designed to do fewer tasks but they are sent more frequently, thereby helping to lessen the likelihood of equipment failure.

In 1997 the first pair of probes were sent: the Mars *Global Surveyor* and the Mars *Pathfinder*, an orbiter and lander, respectively. In keeping with NASA's new and more economical policy, the *Global Surveyor* was built from spare parts left over from the *Observer*; *Pathfinder* was kept to a strict three-year streamlined budget, boasting only a limited range rover to survey the immediate area near the landing site. Still, with all the cost-cutting, the *Global Surveyor* and *Pathfinder* have become two of the most successful Mars missions to date.

The first article in this section, "Craft Arrives on Mars and Relays Images," by John Noble Wilford, was published in the *New York Times* the day after the *Pathfinder*'s landing on July 4, 1997. It is included here to demonstrate the initial euphoria surrounding the successful landing. The next article, William Newcott's "Return to Mars," published in *National Geographic* a year later, presents an overview of the mission as well as describes the amount of excitement that it had generated. This second piece also looks to the future of Mars exploration and details what the reader might expect from NASA over the next decade or so. "Revealing the Secrets of Mars," by Nadine G. Barlow for *Ad Astra* discusses these missions, as well as the two missions currently on their way to Mars: the Mars *Climate Orbiter*, launched in December 1998, and the Mars *Polar Lander*, launched in January 1999. The former of these missions is set to arrive in Mars orbit in September 1999 and will provide command and data relay support for the Mars *Polar Lander*. The *Polar Lander* will arrive on Mars in December 1999, and as its name suggest, will land at the Martian south pole and dig into the soil to take samples for study. In the final article of this section, "NASA Still Dreams of Building an Outpost for People on Mars," Warren E. Leary of the *New York Times* reports on NASA's quiet preparations for a manned mission to Mars. Administrators hope that as they look into cheaper and safer ways to carry out such a mission, it might become more feasible economically and politically to sell the idea to the president and congress. "The space station is only justified by the idea of a trip to Mars," Dr. Louis Friedman, executive director of the Planetary Society, a group that advocates human space travel, told Leary, "Why study the biological effects of long-term space flight on people, and a lot of other technology, if you are not going to go somewhere?"

Craft Arrives on Mars and Relays Images[1]

PASADENA, Calif., July 4—After a smooth seven-month journey, an American spacecraft came to a bouncing, rolling landing on Mars today to begin intensive studies of what the planet is made of and whether life could have possibly arisen there in its distant past.

The first landing on Mars in 21 years, which mission officials described as nearly perfect, occurred shortly before 10 A.M. here in California (1 P.M. Eastern time). It was still night on the dusty, rocky red plain of Ares Vallis, dry and freezing, almost exactly where Mars *Pathfinder* had been aiming. Almost immediately, radio signals indicated a successful arrival.

"I'm ecstatic, absolutely ecstatic," Brian Muirhead, deputy project manager here at the Jet Propulsion Laboratory, exclaimed after the first data came streaming in. "We are on the surface of Mars and have received our first telemetry."

More than six hours later, the first transmitted photographs showed the spacecraft and its attached roving vehicle in the foreground, resting upright, as well as a broad field of rocks and sand stretching to hills on the far horizon. A color mosaic of 120 of these pictures was made public tonight.

But an examination of the pictures revealed a hitch that project officials said would almost certainly delay the deployment of the six-wheel rover, named *Sojourner* and weighing 23 pounds on Mars, until late Saturday afternoon. Such a delay, scientists said, was not surprising and should not reduce the mission's research output.

Looking at the pictures, engineers saw that the air bags that had cushioned the spacecraft's impact were deflated, as planned, but some of them were bunched up near the rover. To retract the bags more fully, commands were sent to the craft to lift the petal-like part of its outer shell where the rover is still stowed. While the petal is lifted, electric motors will try to pull the air bags away from the rover.

Until this is done, engineers plan to hold off releasing the two ramps down which the rover is supposed to roll onto the Martian surface. *Sojourner* will be the first mobile vehicle to operate on another planet and is a test model for more advanced roving vehicles to be used on future missions to Mars.

Sojourner will be the first mobile vehicle to operate on another planet and is a test model for more advanced roving vehicles to be used on future missions to Mars.

While controllers waited for *Pathfinder*'s first pictures, Vice President Al Gore telephoned his congratulations to Daniel S. Goldin, Administrator of the National Aeronautics and Space Administration, who was here for the landing, and to the engineers and scientists of the project.

President Clinton said in a statement that the return to Mars "marks the beginning of a new era in the nation's space exploration program."

At each step in Mars *Pathfinder*'s descent to the surface and early operations, cheers broke out in the control rooms here, cheers that carried echoes of a collective sigh of relief, for the record of flying to Mars has been spotted with frustrating failure.

Over the years Russian spacecraft have suffered seven disastrous malfunctions while scoring only four partial successes. Until now, the United States had six successes, beginning with the *Mariner 4* fly-by of the planet in 1965, and three failures. The most recent of these came in 1992, when Mars *Observer* was inexplicably lost as it approached the planet for a planned orbiting mission.

There was relief also because much is riding on *Pathfinder*'s success. The mission is not only the first in an ambitious program of Mars exploration in the next decade but also a critical test of a new approach to planetary expeditions by the United States. Instead of mounting a few complex, multipurpose and expensive missions, the emphasis now is on building several relatively low-cost craft with limited objectives that can be ready to go to Mars every 26 months.

Dr. Wesley T. Huntress Jr., the chief of space sciences at NASA, said this "amounts to the second era in the exploration of Mars."

As the first such craft, *Pathfinder* was launched three years after development began, and the entire mission is estimated to cost $266 million. The last American mission, the failed *Mars Observer*, cost $1 billion. A second low-cost craft, *Mars Global Surveyor*, is on its way to orbit Mars in September for a two-year mapping reconnaissance, which had been Mars *Observer*'s primary objective.

NASA's plans call for two missions—a lander and an orbiter—to be flown to the planet at each of the subsequent 26-month windows of opportunity, culminating in a mission in 2005 to bring back samples of carefully selected Martian rocks and soil for analysis. Scientists think there is little hope of answering life-on-Mars questions until they can study those rocks.

Although *Pathfinder* and its rover are not equipped to search for signs of life, project scientists expect a device on the rover called an alpha proton X-ray spectrometer to determine the mineral content of rocks scattered several hundred feet from the landing site. Since the rocks are thought to have been washed onto the plain from the ancient highlands nearby, their mineral composition should provide clues as to the planet's early environment, whether it was indeed once a warmer, wetter place where microbial life could have possibly developed.

The Ares Vallis landing site was selected by geologists, who studied photographs taken by the Viking orbiters, because it appears to be a flood plain where a variety of ancient highland rocks should have been deposited after catastrophic floods. The site is about 19 degrees north of the Mars Equator and some 500 miles southeast of the place where the first Viking spacecraft landed in July 1976. That was the first successful landing on the planet.

The two Viking landers conducted life-detection experiments, but failed to detect any indisputable signs of biological activity. That result dampened lingering hopes of finding Martian life and interest in continued exploration of the planet, until geologists in recent years pointed out that the many eroded channels on Mars indicated that liquid water once flowed over the surface and just might have created conditions supporting life.

After plans for Mars Pathfinder were already under way, NASA and university geochemists announced last August the discovery of minerals, hydrocarbons and possible microfossils in a meteorite that fell from Mars. This evidence, they said, indicated that life might have existed on Mars in its first billion years. Their interpretation, though controversial, excited new interest in exploring Mars.

Dr. David S. McKay, a leader of the meteorite analysis team at the Johnson Space Center in Houston, said today: "We think we have some additional lines of supporting evidence. But we do not have the smoking gun yet."

Dr. McKay said he doubted that Pathfinder would find anything to corroborate the meteorite findings.

"The best we can hope for is that the rover will perhaps find interesting rock types, sedimentary rocks," he said, referring to the kind of rocks usually formed in the presence of water. "Beyond that, if we could actually see a fossil in a rock, but I think that's not likely."

The Pathfinder landing craft is equipped with instruments for recording weather conditions and a camera with two lenses and 24 filters for producing color stereoscopic photographs of the landscape. The camera can swivel 360 degrees to take panoramic shots.

NASA . . . announced last August the discovery of minerals, hydrocarbons and possible microfossils in a meteorite that fell from Mars. This evidence, they said, indicated that life might have existed on Mars in its first billion years.

"We will have a lot of opportunities to make tremendous discoveries," said Dr. Peter Smith of the University of Arizona, who led the team developing the camera system.

The rover also carries three color cameras, two forward and one rear, to help it navigate the surface and take close-up pictures of the rocks it examines with its X-ray spectrometer. As planned, the rover should be able to nose up and press the spectrometer to the surface of a rock, zapping it with protons. An analysis of the backscattering protons should tell geologists the nature and composition of the rock. The instrument is sensitive to all chemical elements except hydrogen and helium.

For a few anxious hours today, though, all that mattered was getting *Pathfinder* to the surface in working order.

Navigation experts reported the spacecraft's aim was true as it raced toward the planet, accelerating to 16,600 miles an hour because of the pull of Martian gravity. The craft was operating autonomously, by commands stored in its computer. Given the 10 1/2 minutes it takes for radio signals to travel from Mars to Earth, there was no time for ground controllers to direct each event or respond immediately to problems.

At about 5,300 miles above the surface, the spacecraft jettisoned its cruise stage, which contained fuel, rockets and other systems for maneuvers over the seven-month flight. Half an hour was then left before the craft plunged into the fringes of the thin Martian atmosphere, which is mostly carbon dioxide.

The atmospheric friction should have begun slowing the spacecraft, at 80 miles above the surface and heating its shielding to a glowing red. This was an unseen spectacle in the dark of the Martian night. But flight controllers, monitoring a faint radio signal, could tell that *Pathfinder* was indeed diving through the atmosphere and decelerating.

"Spacecraft is now slowing down very rapidly," announced Rob Manning, the chief engineer.

At maximum deceleration, engineers figured, the spacecraft was being jolted by forces nearly 20 times greater than Earth's gravity, much stronger than anything felt by astronauts re-entering Earth's atmosphere.

But was *Pathfinder* coming in at a safe angle? Its entry into the atmosphere had been at an angle of about 14 degrees. Any sharper, and the craft would probably burn up. Any shallower, and it would skip out of the atmosphere and soar off into space.

At an altitude of less than seven miles, the parachute unfurled to another round of cheers in the control room.

Drifting down, the craft released its heat shield, then detected the ground with its radar, at one mile up. Then the air bags inflated, descent rockets fired for a few seconds and *Pathfinder* dropped to the surface at a velocity of 23 miles an hour, bouncing at least three times.

The entry and descent had taken four and a half minutes. At first, there was only radio silence from *Pathfinder*. Then, at 10:08 A.M., Mr. Manning reported a barely detectable signal, which he said was "a very good sign."

One by one, signals reassured controllers that the spacecraft had apparently landed upright, that the deflated air-bags had been retracted and that the outer shell had opened to expose the camera and rover.

At this, Mr. Manning, who supervised the entry operations, announced facetiously, "This is Rob Manning, and I'm out of a job."

Dr. Edward C. Stone, the director of the Jet Propulsion Laboratory, declared, "It certainly appears we have landed and have a good operating system."

By then, the sun was rising on Mars *Pathfinder* and its landing site at Ares Vallis.

Return to Mars[2]

The midsummer sun was high in a clear yellow-brown sky. The morning's filmy blue clouds had dissipated, and the temperature was eight degrees Farhenheit—way up from last night's low of minus 100 degrees. A breeze wafted from the west at about eight miles an hour.

A perfect afternoon for a drive on Mars.

Gingerly pushing a joystick, I watched the computer screen as the six-wheeled rover named Sojourner eased away from the Mars *Pathfinder* lander, which had carried it to this rocky Martian plain 119 million miles from Earth. The two-foot-long vehicle rolled along a Mars landscape replicated from images beamed back to Earth after *Pathfinder*'s landing on July 4, 1997.

I was not, of course, commanding the real rover. By the time I sat down at this computer terminal at the Jet Propulsion Laboratory (JPL) in Pasadena, California, *Pathfinder* had been out of contact for weeks. But if the lander had still been alive, I could have been plotting real rover maneuvers at this computer.

An intriguing collection of rocks lay a few yards to the left of the rover, but a pair of good-size stones seemed to block my path. Luckily, I was wearing liquid-crystal 3-D glasses that enabled me to see depth on the flat computer monitor. There would be just enough room for me to squeeze *Sojourner* through—I thought.

"Um, you may have a problem there," said a charitable Brian Cooper, who designed this virtual-reality computer program for the Jet Propulsion Laboratory and served as the primary designated driver for the actual rover's three-month mission. From this very console in his JPL office, the bearded, shirt-sleeved Cooper plotted the moves of the interplanetary dune buggy.

Fortunately for the U.S. space program, my navigational mishap happened off-line. As opposed to my blunderings (somehow I found a way to make the virtual rover rise into the air and fly off into the distance, growing ever tinier as it disappeared over the Martian horizon), Cooper found routes around barriers, stopped to spin the rover's wheels so scientists could study the soils stirred up, and cozied up to rocks for closer looks.

"Don't feel too bad," Cooper told me. "We spent so much time in one area that I nicknamed it the Bermuda Triangle."

With the goggles and video-game graphics, this all seemed like entirely too much fun for real science. In fact, from the beginning there was a vague sense of goofy abandon to the Pathfinder project. The spacecraft was designed, built, and launched in three years. The mission's total cost ran 265 million dollars, one-fourteenth the amount of the last successful Mars missions, Viking 1 and 2, in 1976. The rover was so cute that a copy became one of the most popular Hot Wheels toys ever produced. Even the landing was offbeat: At 1,165 feet above the surface, aher being slowed by the atmosphere and a parachute, the lander sprouted multiple air bags, cushioning itself inside a huge beach ball. It bounced more than 15 times across the Martian surface before rolling to rest on a gentle slope.

The air bags were deflated and cranked back around the lander, which then unfolded its petal-like shields to reveal the payload: the Imager for Mars *Pathfinder*—(IMP), a stereoscopic camera with 24 filters; the Atmospheric Structure Instrument/Meteorology (ASI/MET) package to record daily weather; and the rover itself, with cameras on the front and back and an alpha-proton x-ray spectrometer (APXS) to analyze the makeup of Mars rocks and soil.

That may seem like an ambitious lineup, and those devices did come up with some remarkable discoveries. But at its inception the main aim of the entire Pathfinder mission was simply to get something—anything—safely on Mars.

"This was primarily an entry, descent, landing demonstration said Matthew Golombek, the project scientist whose wide-eyed smile and unbridled enthusiasm endeared him to TV viewers as Pathfinder's spokesman. "After we had the thing on the surface, whatever we did was pretty much considered gravy."

From his windowless little office on the JPL campus, Golombek oversaw the Pathfinder science mission. "They probably let me have the job thinking I wouldn't cause them too much trouble," he smiled. But I thought it wouid be silly just to send a red brick to Mars. I figured we should try to learn something new."

The learning began even before the landing. During the descent the spacecraft recorded information about the planet's atmosphere.

Finally on the ground in a region called Ares Vallis, the IMP camera showed rocks—large and small, angular and rounded, dark and bright—stretching to the horizon. To Golombek the landscape was gratifyingly familiar.

The mission's total cost ran 265 million dollars, one-fourteenth the amount of the last successful Mars missions, Viking 1 and 2, in 1976.

"I spent two and a half years worrying about the landing site," he said. "We knew from old *Viking* images that there were areas of Mars that looked like they had been formed by catastrophic floods. On Earth places like that are where you can get a variety of rocks, so that was where we wanted to go."

As 3-D images of the surface were processed, it became clear that some kind of trough lay just beyond a nearby ridge. Set on the trough's edge like books on a shelf was a collection of angled rocks. The consensus grew that what carved the trough and deposited many of the rocks at *Pathfinder*'s landing site was a flood whose volume may have equaled that of all the Great Lakes.

And when the rover ventured off the lander, it saw rocks that are possibly conglomerates, a type of rock that forms over millennia as water rounds pebbles and cobbles and deposits them in a matrix of sand and clay. "That means there was once liquid water on Mars," said Golombek. "It suggests a very different climate, perhaps one where life could have developed. That raises the questions: If life developed, what happened to it . . . and if not, why not?"

"If life developed, what happened to it . . . and if not, why not?"

From the earliest times humans knew there was something different about the bright red heavenly body that marched across the night sky out of step with the steady progression of the stars. The ancient Sumerians, Greeks, and Romans associated it with their god of war, unaware that its blood-red color was merely evidence of a world covered with iron oxide dust.

In the late 1800s an Italian astronomer named Giovanni Schiaparelli mapped what he described as *canali* on the Martian surface. The word can mean either natural channels or artificial canals, and Percival Lowell, a Boston millionaire and astronomy enthusiast, seized on the more tantalizing translation. He decided he would see those Martian canals for himself. And he spared no expense.

Lowell called the Arizona mesa on which he established his observatory Mars Hill. Back in 1894 it was in the middle of nowhere. Today it's a five-minute drive from where old Route 66 passes through Flagstaff. Stroll up a pathway—past the domed mausoleum where Mr. Lowell's remains lie—and you're at the door of the wooden cylindrical building that houses the 24-inch refracting telescope he brought to the site by train.

Lowell spent years squinting through the telescope, drawing the intricate patterns he saw—or imagined he saw—on the Martian surface. The canals, he said, stretched from the cold ice caps to regions closer to the equator. It was clear to

Lowell that Mars was a drying, dying planet and that its ingenious inhabitants had created the canals in a last-ditch attempt to survive.

It was, needless to say, one of the great wrong guesses in astronomical history. But Lowell helped propel a century of Mars studies—along with classic science-fiction books and movies that planted the seed of Martian curiosity in the young minds of more than a few future scientists.

Science fiction aside, Mars is in many ways remarkably like Earth. A day on Mars lasts 24 hours, 37 minutes. The Earth's axis tilts at 23.45 degrees; Mars's tilts at 25.19 degrees. Both planets have observable seasons, with warm summers that melt their polar ice caps. Clouds drift across the face of each. Mars is half again as far from the Sun as Earth, yet while polar nights are nearly minus 200 degrees Farhenheit summer days south of the equator can get as hot as 80 degrees.

Although the diameter of Mars is little more than half that of Earth, its major geological features dwarf those on our planet. The Martian volcano Olympus Mons rises 75,000 feet, two and a half times the height of Everest. And the Valles Marineris canyon, which would stretch from San Francisco to New York, is the longest such valley known in the solar system.

The more Earthlings learned about Mars, the less it remained an astronomical curiousity. It was a place to go.

But you wouldn't want to live there. As Elton John sang in the 1970s, "Mars ain't the kind of place to raise your kids / in fact it's cold as hell." Dusty, pockmarked, dead: This was the view of Mars sent back to Earth by the Mariner flyby missions of the 1960s. The black-and-white images showed a planet that looked positively moonlike. Impact craters were everywhere. Then, in the early 1970s, the *Mariner 9* orbiter showed intriguing evidence of volcanism and dry riverlike flood channels. Two 1976 landers, *Viking 1* and *2*, looked for life on Mars by scooping up and testing soil samples, which yielded no organic material.

For nearly two decades after the *Vikings*, Mars remained unvisited by the United States. (The Soviet spacecraft *Phobos 2* orbited the planet for a month in 1989.) A grand return was planned with the Mars Observer probe, which bristled with experiments, sensors, and cameras. A week before its planned arrival in August 1993 a scientist showed me a single black-and-white image of Mars taken by *Observer* to test the equipment.

"If anything happens to that spaceship," I told him "this will be the most expensive eight-by-ten glossy ever taken." We laughed. Days later *Observer* fell silent.

The media sensation Pathfinder caused was unlike that of any space program since Apollo. In the first month of surface operations the JPL Mars Pathfinder Internet site registered an unprecedented 566 million hits.

The Mars *Observer* debacle came at the end of an era: The age of spending a decade or more developing, building, and launching a space probe was over. NASA's chief, Dan Goldin, summarized the future of space exploration in three words: "Faster, better, cheaper."

NASA announced that spare parts for *Observer* would be assembled on a new craft called *Mars Global Surveyor*. It would be launched as soon as possible, the first in a series of Mars orbiters, landers, and rovers that included *Pathfinder*.

With its three-year deadline and strict budget, the design and assembly of *Pathfinder* was unlike any space project before it. In the old days a device as critical and complex as the IMP camera would have been built entirely by a government contractor. But now, to save money, the final assembly and testing was done by scientists, engineers, and graduate students at the University of Arizona Lunar and Planetary Laboratory in Tucson.

"In the past, large teams were common on this kind of project," said Peter Smith, the imaging team leader who conceived the camera at the University of Arizona, helped build it, and then supervised its operation on Mars. "This had no more than 20 employees. Traditionally with contractors the cost of the camera would have run in the tens of millions of dollars. Ours cost less than six million."

Tall, bearded, and obviously more comfortable in knit polo shirts than anything else, Smith recalled the hectic last days of building the camera in December 1995.

"The motors for pointing the camera arrived here at the university on December 20, and the camera was due at JPL December 29. That was the Christmas from hell, but we worked very long hours and got the thing packaged at 4:00 A.M. on the 29th.

"Before sending the camera to JPL, my colleagues Chris Shinohara and Bob Marcialis and I put it on my dining room table and toasted it with warm beer, the only thing we could find at five in the morning. That was real team spirit."

The Global Surveyor orbiter was launched first, in November 1996. *Pathfinder* was hurled into space a month later, on a trajectory that brought it to Mars two months sooner. The media sensation Pathfinder caused was unlike that of any space program since Apollo. In the first month of surface operations the JPL Mars Pathfinder Internet site registered an unprecedented 566 million hits.

In a JPL conference room a blowup of the Martian panorama surrounding the *Pathfinder* lander stretched nearly wall to wall.

Almost obliterating the image were over a hundred yellow adhesive Post-its bearing the whimsical descriptive names that scientists had given Martian rocks. Names like Barnacle Bill, Yogi, and Couch.

"It seemed like a better idea than just assigning them numbers," said the young scientist showing me around. "But they were careful not to give them the names of any real people. Too much danger of jealousy."

At that moment my eye caught sight of a rock named Moe, so named, it appeared, for its shaggy "haircut." A momentary lapse, I supposed. And so it came to pass that the only feature the mission named for a person honors the memory of Moe Howard, the eye-poking, skillet-wielding leader of the Three Stooges.

Each day during the mission Brian Cooper, the rover driver, sat at his computer screen and mapped out *Sojourner*'s route, a trail the rover followed at a blistering two feet a minute. Cooper couldn't simply rev it up and drive it around like a kid with a remote-control car though. Each new route had to be painstakingly planned and tested by the rover team. That's why Cooper had to develop a computer program that enabled him to envision the rover in the Mars landscape. He rehearsed the day's intended route at JPL, then sent the set of commands, the radio signals taking some 10 minutes to reach Mars. He gave the rover way points, then let it head for the destination on its own, using five lasers and two cameras for range finding and to see what was in its path.

"The cameras and lasers worked as an avoidance system," he said." Even if I told the rover to go off a cliff, it wouldn't do it."

Cooper watched me fumble with his baby. Finally, a confession. "I think this is the most fun job on the whole project," he said. "I've had a blast."

Cooper's most delicate procedure was to maneuver close enough to rocks for the rover's APXS sensor, at the end of a short movable arm, to be pressed against them.

"We found rocks very high in silicon. which indicates that some crustal materials are like the continental crust on Earth," said Matthew Golombek, the project scientist.

The *Pathfinder* lander transmitted its last data on September 27. Though twice revived by JPL scientists, it sent no further information. Wild Martian temperature changes probably caused a wire to snap or a soldering point to crack. The solar-powered rover, however, may still be rolling along using its laser eyes to dodge rocks as it circles the lander like an orphaned pup. More likely, the rover has at some point sensed itself in a precarious position and placed itself on hold, waiting for instructions that will never come.

By September 27, the rover had covered about 110 yards of terrain and taken 16 APXS readings of Mars rocks and soil. The lander and rover sent back more than 17,000 images. Almost daily weather recordings tracked temperature, air pressure, and wind speed and direction, including several small, swirling dust devils passing right over the lander. "In short," said Golombek, "we have explored about 240 square yards of Mars."

Like most Pathfinder scientists, Golombek keeps coming back to the mounting evidence that Mars, like present-day Earth, once had bodies of water lapping against the now dry landscape.

Even in its dryness Mars betrays a watery past. All around *Pathfinder* were windblown dune forms. Rocks visible from the lander had been sculpted by wind, a phenomenon that requires airborne sand as an abrasive. On Earth, you need running water to make sand.

And there were more signs of water on Mars. One of the experiments added to the *Pathfinder* payload was a series of magnets on the lander. As the weeks on Mars wore on, the IMP camera saw dust collecting around the magnets. The patterns confirmed that the particles, just two microns across, were highly magnetic—interpreted as evidence that iron in the crust was once leached out by groundwater.

Soon after *Pathfinder*'s last transmission, *Mars Global Surveyor* began sending back remarkably high-resolution images

Mars Exploration:

- Late 1998–early 1999: Mars Climate Orbiter will carry cameras and equipment to study the atmosphere and the surface. Mars Polar Lander will settle near Mars's southern ice cap-thought to be frozen water and carbon dioxide— and take samples with a robotic arm. Two small probes will be dropped into the ground to search for water. No rover will be included.

- March 1999: Mars Global Surveyor will attain a circular orbit over Mars's poles. Capable of imaging objects the size of a large desk, it will map the Martian surface. Other instruments will continue to study Mars's localized magnetic fields. A laser altimeter will measure, to an accuracy within a meter, seasonal changes in the height of the ice caps.

of the Martian surface. As it orbits Mars, *Global Surveyor* is using the atmosphere to slow itself down until it attains an ideal operating orbit in March 1999. Meanwhile, its near–spy-satellite-quality camera has revealed that the walls of Valles Marineris have sharply defined layers, like the Grand Canyon. The orbiter's laser altimeter, which measures the distance from the satellite to the planet's surface, appears to show that the north polar cap rises much higher than previously thought, in places almost a mile above the relatively flat, sandy plains that surround it. The laser altimeter also shows a vast flat region covering most of Mars's northern hemisphere—possibly an extinct seabed or ancient mudflat.

Said Golombek, "It just seems everywhere you look on Mars, you see water. At least the evidence of it."

So where did it go? The prevailing view holds that most of Mars's water is frozen—at the poles, underground, or on the planet's northern plains. But life cannot exist without liquid water. No water . . . no Martian life.

Ian McCleese got hooked on Mars early. As a 16-year-old in San Diego, his high school science project was to try to grow slime mold in a simulated Mars environment. Young McCleese had a man at a local garage weld together a chamber into which he pumped carbon dioxide and traces of oxygen. The temperature was kept low by immersing the chamber in a refrigerant bath. A vacuum pump lowered the air pressure, and an ultraviolet lamp simulated sunrays unfiltered by a thick atmosphere.

"I killed the slime mold," he reports 30 years later.

A Timetable

- 2001: The new millennium begins with the launch of a new orbiter and a lander, ideally with a rover payload. Considerably hardier than the fragile *Sojourner*, the 2001 rover will collect rocks and soil samples and cache them. Scientists will select the landing site based on high-definition data collected by Mars Global Surveyor and Mars Climate Orbiter.
- 2003: As in the 2001 mission, this lander will be sent to an area with rocks most likely to bear evidence of past life on Mars. Again, thee rover will collect and cache rocks and soil samples.
- 2005: The first round-trip mission to collect rocks. Using technology not yet developed, a lander will head for the most promising of the previous landing sites. Its rover will collect the rocks and soil samples cached by an old rover. Samples will be brought back to Earth in 2008 for detailed study.

Despite that, McCleese is the chief scientist for JPL's Mars Exploration Program, setting the strategy for an ambitious series of Mars missions that began with *Pathfinder* and *Mars Global Surveyor.*

"We'll be going back every 26 months, each time Mars is at its optimum position with respect to Earth," he said. Pairs of spacecraft—one with an orbiter, the other with a lander—will be launched on separate rockets to the red planet. Besides carrying their own scientific equipment, the orbiters will act as relay stations transmitting data back to Earth from the landers.

"I plan to be here through all the planned missions and for subsequent sample-return missions as well," said McCleese. "We really don't believe a single sampling will tell us everything we need to know about Mars."

Already completing his camera for *Mars Polar Lander,* Peter Smith both revels in his work and longs for something more. Wearing a pair of blue-and-red 3-D glasses, he studied a stereo panorama of Mars mounted on a wall of the IPL auditorium.

"If I have a regret, it's that we lost *Pathfinder* just before the Martian dust storm season," he said. "I tell you, I would have loved to see a big old cloud of dust roiling up from that horizon. It would have been like the Dust Bowl, only a whole lot bigger, covering the entire planet."

And, presumably, louder. Perhaps we'll find out in 1999, when *Mars Polar Lander* settles in near the South Pole—with a microphone on board.

Smith stepped closer to the mural for a look at the panorama's most distinctive features, two hills he named Twin Peaks.

"They're a half mile away, and we couldn't get there with the rover. But our next generation rover will be able to drive beyond the horizon, leaving the lander behind.

"We can't afford to launch a heavy rover that can drive hundreds of miles, however, so the irony is that when you go to Mars cheaper and faster, it actually takes longer to really explore it. I'd love to see a Mars program that would get humans there in our lifetime."

In a fit of optimism, then President George Bush suggested that the U.S. should land humans on Mars by 2019, the 50th anniversary of Neil Armstrong's first step on the moon. As the years pass, that possibility seems to dwindle. But just imagine the year 2008. The sample mission comes back with a rock that bears evidence of fossil remains.

There may have been life on Mars. The only way to know for sure would be to go look.

We'd still have 11 years. We got to the moon in less than 10.

Revealing the Secrets of Mars[3]

It is an exciting time to be an aficionado of Mars. From the stunning success of last summer's Mars Pathfinder mission and the extraordinary surface views being provided by *Mars Global Surveyor* to the series of upcoming orbiter and lander missions designed to address the climatic evolution and the possibility of past life, many of the secrets that Mars has sheltered are being revealed for the first time. Although currently a cold, dry world, Mars in the past has apparently been more Earth-like with a thicker atmosphere, warmer temperatures, substantial quantities of surface water, and perhaps even early stages of life. NASA's 10-year program of dedicated Mars missions promises to reveal more surprises about the evolution of our neighboring world.

Although currently a cold, dry world, Mars in the past has apparently been more Earth-like with a thicker atmosphere, warmer temperatures, substantial quantities of surface water, and perhaps even early stages of life.

Mars Global Surveyor Mission

With the demise of Mars *Pathfinder* on the Martian surface last September, the Mars Global Surveyor (MGS) mission is currently the only operating spacecraft at Mars. This mission is a replacement for the Mars Observer Mission (MOM), which was lost during Mars orbit insertion in 1993. MGS carries most of the same instruments as MOM and its scientific mission is the same: to characterize the surface properties and climatic variations over an entire Martian year (687 Earth days) from a circular orbit. MGS was launched on 7 November 1996, aboard a Delta II rocket from Cape Canaveral Air Station, FL. The three-axis stabilized spacecraft achieved Mars orbit insertion on 11 September 1997, and since that time has been undergoing aerobraking to reduce its elliptical orbit to a circular mapping orbit. Two solar arrays (each 3.5 x 1.9 m in size and providing 980 watts of power) are being dipped into the Martian atmosphere to provide the drag for the aerobraking. Original plans called for the circularization of the orbit to be completed in early 1998, but structural damage on one of the solar panels caused engineers to redesign the aerobraking sequence so that the pressure exerted on the solar panels is only one-third that originally planned. As a result, the circular orbit will not be achieved until March 1999. The final circular orbit will be at an altitude of 378 km, near polar (inclination of 92.5 degrees), and sun synchronous at the 2 A.M. / 2 P.M. position.

---

3. Article by Nadine G. Barlow. From *Ad Astra* p22–26 July/Aug. 1998. Copyright © 1998 National Space Society. Reprinted with permission.

The instruments aboard MGS arc designed to study various aspects of the Martian environment. The Mars Orbiter Camera (MOC) has a narrow-angle high resolution camera (providing images with resolutions down to 1.5 m per pixel) and two wide angle cameras (down to 230 m/pixel resolution) to study the Martian atmosphere and surface. The Thermal Emission Spectrometer (TES) studies the Martian atmosphere and surface using thermal infrared spectroscopy, providing information on the composition and particle size of surface materials and atmospheric dust. The Mars Orbiter Laser Altimeter (MOLA) uses a neodymiumdoped yurium aluminum garnet laser transmitter to bounce laser pulses off the surface of Mars to determine the global topography. The Radio Science (RS) measurements use the spacecraft's radio system to study the gravity field (which contains information about mass distributions both above and below the surface) and the structure of the atmosphere. The magnetic experiments consist of two magnetometers (MAG) to establish the nature of any magnetic field and an Electron Reflectrometer (ER) to map remnant magnetic fields retained by crustal rocks. This suite of instruments allows many questions about the Martian environment to be addressed in levels of detail previously unavailable to planetary scientists.

Early Results From MGS

Although MGS is still in its aerobraking period some data have been collected and preliminary results are in. One of the outstanding questions about Mars has been whether it has an active magnetic field. Earlier Mars and Mariner spacecraft results were inconclusive as to whether a weak active field was present or if the results could be explained entirely by either interaction of the solar wind with Mars' ionosphere or remnant magnetism retained in crustal rocks. The MGS MAG/ER experiments have not detected any significant active field but crustal magnetic anomalies have been detected, indicating that Mars has had a magnetic field in its past. Currently the data coverage is too sparse to determine the global extent of the crustal magnetic anomalies and thus reliable estimates of the time period during which the magnetic field was active are not yet available.

Over 200 profiles of the vertical structure of the upper Martian atmosphere (thermosphere) have been obtained by MGS, compared to three obtained by previous missions (Mars *Pathfinder* and the two Viking Landers). These profiles provide information about density, temperature, and pressure with altitude. Seasonal and diurnal variations are seen between the MGS density profiles and those from the previous missions. Thermospheric storms, at least one associated with a dust storm, have been detected from increases in density, and thermospheric density bulges have been detected at 90 degrees W and 270 degrees W longitude, corresponding to topographic highs on the planet's surface. These results will help scientists to better understand the global circulation of the planet.

Another major advance in our understanding of Mars has come from analysis of the MOLA topographic data. Although relative topographic variations have been known since 1972 from *Mariner 9* data, the detailed topography needed to understand many of the features on Mars is only now being provided by MGS. Even with the present elliptical orbit, MOLA is providing vertical resolutions of about 30 cm with horizontal resolutions of 300 to 400 m. MOLA has been able to provide detailed topographic information about individual features such as impact craters volcanoes, fractures, channels, and polar deposits. One discovery is that some of the channels, including Ares Valles in whose outwash area Mars *Pathfinder* landed, are deeper than previously thought, indicating more water has flowed through the channels than earlier suspected. In addition, MOLA has revealed that the northern plains of Mars are extremely flat, as smooth as the Earth's oceanic abyssal plains. The smoothness of the northern plains supports the theory that they are sediments deposited in a vast ocean which once covered this area.

The goals of TES are to determine compositions and particle sizes of surface materials and atmospheric dust, study the polar cap deposits, determine thermophysical properties of surface materials, locate ice clouds and determine their temperatures, and characterize the thermal structure and dynamics of the atmosphere. TES has confirmed observations from previous missions that dust is a large component of the Martian atmosphere and that water-rich clouds are found at altitudes up to 55 km. TES has detected several dust storms that have occurred since MGS entered orbit, including a major dust storm over the Noachis region on 25 November. Bright surface deposits appear to be composed primarily of pyroxene, consistent with Earth-based remote sensing results. Upper limits can be placed on the concentrations of certain minerals, including carbonates (less than 10 percent), olivine (les than 10 percent) clay minerals (less than 20 percent), and quartz (less than 5 percent) in the limited regions observed. TES also has shown that small dust storms are common along the edges of the south seasonal polar cap as the cap begins to retreat in size.

The most visually exciting results from MGS come from the cameras. High resolution images from MOC show ubiquitous

Some of the channels, including Ares Valles in whose outwash area Mars Pathfinder landed, are deeper than previously thought, indicating more water has flowed through the channels than earlier suspected.

evidence of dust and sand on the surface, and reveal how these materials have been manipulated by the Martian winds. Dunes and drifts are seen almost everywhere, regardless of composition or age, and sediment thicknesses of a meter or more are common. Complex juxtaposition of fresh and old dune fields indicate that prevalent wind directions have changed over the planet's history, perhaps in connection with long-term oscillations in the planet's obliquity and orbital parameters. MOC images of impact crates and the Valles Marineris canyon system have revealed meter-scale layering in the walls, indicating that the upper crust of Mars is highly stratified into layers of varying composition. High resolution imagery of the south polar region reveal the details of the residual ice cap, the layered terrain, and intersecting ridges which may be remnants of older underlying deposits. MOC has also targeted specific regions of interest, including the infamous "Face on Mars" and the landing sites of Mars *Pathfinder* and the two *Viking* landers. The wide angle cameras have also been used to monitor ice clouds over the Tharsis volcanoes and dust storms in the southern hemisphere.

In less than a year after its arrival at Mars, MCS has already dramatically changed our perspective of our neighboring world. Even bigger surprises are almost certain to come from the upcoming Mars missions that NASA has planned.

Upcoming Revelations—Volatiles, Climate, and Life

More secrets are expected to be revealed as a series of upcoming orbiters and landers focus on specific questions about Martian climate and evolution. The next spacecraft in the NASA Mars queue is the Mars Surveyor 98 mission, which consists of the *Mars Climate Orbiter* (MCO) and the *Mars Polar Lander* (MPL). The two sections launch separately: MCO has a 10–23 December 1998 launch window while the MPL will be launched sometime between 3 and 16 January 1999. MCO will arrive at Mars in September 1999, and be aerobraked into an elliptical orbit with a periapse altitude of 160 km. The orbiter will provide command and data relay support during MPL's surface operation period, then begin studies of the daily atmospheric structure variations and imaging of the surface using the Mars Color Imager. The mapping mission will extend for one Mars year (687 days), followed by up to three years of orbiter support as a communication relay for future Mars landers.

The MPL arrives at Mars in December 1999, and descends to the surface on a direct trajectory (i.e., no descent from a Mars orbit). The landing site is between 74 degrees S and 78 degrees S latitude, within the polar layered deposits and less than 1000 km from the South Pole. Shortly before atmospheric entry, two attached microprobes will detach from the lander, separately enter the atmosphere, and hit the ground at about 400 mph. The forebody of each probe will penetrate up to six feet into the Martian soil, providing information on target properties. The microprobes also have a small drill to sample the subsurface soil and determine if ice is present. Measurements of the soil temperature and the atmospheric pressure will be made every hour for at least two Mars days. Meanwhile, MPL will have landed and begun collecting data using the *Mars Volatiles Climate Surveyor* (MVACS) instrument package and the LIDAR instrument. MVACS includes meteorology, imaging, and soil composition experiments that will search for near-surface ice and any record of climatic change cycles recorded in the polar layered deposits. LIDAR will detect the presence of dust and ice in the atmosphere using a laser. MPL also includes a robotic arm that will be used to collect soil samples for analysis and test surface properties by digging into the soil. The lander is expected to collect data for about three months.

Unraveling the climatic history of Mars and searching for evidence of possible ancient life on the planet are the goals of the Mars Surveyor 2001 (MSP01) mission. This mission again consists of an orbiter and a lander, and the lander also includcs a rover. The *Mars Surveyor Orbiter* (MSO) is scheduled to launch on 7 March 2001, with arrival at Mars on 10 December 2001. It will be the first spacecraft to utilize aerocapture, using the Martian atmosphere to capture the spacecraft in a circular orbit in one step. MSO will carry two instruments to characterize the mineralogy of the Martian surface: the Gamma Ray Spectrometer and the Thermal Emission lmaging System. The *Mars Surveyor Lander* (MSL) is scheduled for launch on 3 April 2001, with landing on Mars slated for 27 January 2002. Scientists met in January at NASA Ames Research Center to discuss possible landing sites. Although no final decision has been made yet, the landing site will be someplace between 15 degrees S and 5 degrees N latitude and the preference is for an area where water may have been present for long periods in the past, permitting the possible evolution of biologic organisms whose remains could be retained in the rocks and/or soil. MSL instruments include a descent imaging camera, the Mars Environmental Compatibility Assessment (MECA) instrument package that will analyze soil samples obtained by a robotic arm and determine if any components of the Martian soil or dust are hazardous to humans, and the Mars In Situ Propellant Production experiment that will produce rocket propellant using gases from the Martian atmosphere. MSL also will deploy a rover to explore the surrounding terrain. Initial plans called for a long-range rover that would explore for 365 days, collecting rocks and soil for possible later return to Earth. This rover package, called Athena, included a mini-coring device to collect 91 rock and 13 soil samples, plus a Mini-Thermal Emission Spectrometer to characterize the composition of the areas visited by the rover. However, recent reports from Washington indicate that budget constraints will not permit inclusion of the Athena rover and discussion is continuing as to whether the rover will be reduced in size and function or eliminated completely.

If *Athena* is eliminated from the 2001 mission, it may still go to Mars aboard the Mars Surveyor 2003 mission. Plans for this mission are still sketchy, but again would include both an orbiter and a lander. Initial designs call for the lander to include a rover to collect rock and soil samples, similar to the original plans of the 2001 mission. The 2005 mission will then return to either the 2001 or 2003 site, retrieve the samples, and return them to Earth. With the possible changes in the scope of the 2001 mission, the plans for these missions are obviously still in a state of flux. One scenario is that the 2005 mission will land separately from the 2001 and 2003 missions, deploy its own rover to collect samples, then simply return them for analysis on Earth. To stay updated on the plans for these missions, check out the Mars mission Web site at JPL (mars.jpl.nasa.gov). [Editor's note: As of April 12, 1999, this Web site was still valid.]

By 2005, scientists believe we will know enough about the Martian environment to begin educated discussions of sending humans to Mars. We probably will not see humans walking on Mars prior to 2020, but human exploration and eventual settlement of Mars is still a goal for many enthusiasts of the planet. Because of the costs involved, such an undertaking would almost certainly have to be an international effort but enough countries of the world are interested in the exploration of Mars that such a joint effort could become a reality.

Our understanding of Mars has increased dramatically with every mission ever sent to the planet, and the upcoming detailed explorations will continue to astound and surprise us. Mars is an exciting world, one whose evolution has paralleled that of the Earth in some ways but deviated drastically in others. As with all planetary exploration, by better understanding the geologic and climatic evolution of Mars we can gain important insights into our own planet and how to better preserve it for future generations of space enthusiasts.

NASA Still Dreams of Building an Outpost for People on Mars[4]

Washington—As it has through the ages, Mars, the nearest planetary neighbor to the Earth, maintains a strange fascination for people and has long been a coveted destination for the more imaginative.

Whether spurred by science fiction or tantalizing hints of past life in ancient Martian rocks, many people seem to feel that inevitably, humans will someday set foot on the red planet, first to visit, then to stay.

Quietly, and often unnoticed, scientists and engineers are putting together plans and the technology to make such dreams a reality. At several National Aeronautics and Space Administration centers, and at universities and aerospace companies, small-scale studies are under way on differing bits of technology that could come together to send the first human crew to Mars early in the next century.

New approaches being examined for a Mars trip—using lighter, partly inflatable ships, developing closed systems to recycle wastes and produce food, and making rocket fuel on Mars instead of hauling it from Earth—show promise for making the venture safer and more cost-effective, proponents say.

Because there is no political mandate for a human mission to Mars, NASA is approaching the possibility cautiously and with little fanfare. The agency is sponsoring several low-cost research projects aimed at identifying technologies that could be used and readying blueprints for an endeavor, should one be called for.

"I don't know of another event that would inspire our imaginations or stimulate our innate feelings for exploration more than such a journey," said Daniel S. Goldin, the agency administrator. But he emphasized that such planning was tentative.

Mr. Goldin, who has spoken often of his dream that people would one day go to Mars, says that in the next five or six years, he wants his agency to lay the groundwork for such a venture. At the end of that period, he said, criteria for the mission should be solid enough for NASA to show the President or Congress how to accomplish it within 8 to 10 years of setting a starting date.

Small-scale studies are under way on differing bits of technology that could come together to send the first human crew to Mars early in the next century.

---

4. Article by Warren E. Leary. From the *New York Times* F p1 Feb 3, 1998. Copyright © 1998 The New York Times Company.

Four questions must be resolved before there can be a serious human Mars proposal, Mr. Goldin said: Can people live and work in space for the two to four years required? Is there a compelling scientific reason for people to go? Can the journey be made for a relatively low cost? Should the United States conduct the mission alone or as part of an international project?

To address these questions, NASA officials said, the agency has been spending $5 million to $10 million a year on studies directly related to human exploration beyond Earth orbit. Additionally, other projects in the agency's $14 billion annual budget, including research related to the forthcoming International Space Station, could aid a human Mars mission.

Human journeys to the red planet have been proposed before, but they were stymied by their sheer size and enormous projected costs. In 1989, while celebrating the 20th anniversary of the *Apollo 11* Moon landing, President George Bush proposed sending an expedition to Mars in the first decades of the next century. But when NASA and others estimated the cost at $450 billion to $500 billion, the proposal withered.

In 1989 . . . President George Bush proposed sending an expedition to Mars in the first decades of the next century. But when NASA and others estimated the cost at $450 billion to $500 billion, the proposal withered.

"Mr. Bush found very quickly that there was no rationale for a manned Mars mission," said Dr. Louis D. Friedman, executive director of the Planetary Society, a group based in Pasadena, California, that advocates space exploration. "And no one got behind it, neither Congress nor the people."

But things are changing, say Dr. Friedman and other experts. First, the cost estimates for a Mars expedition have dropped sharply in the past few years and should continue to fall as researchers explore new ways of doing things, they say. Recent estimates by NASA engineers put the cost of sending six astronauts to Mars on a two- to three-year mission at about $55 billion, only about 10 percent of the projection a decade ago.

In addition, proponents say, the public is showing a renewed fascination with Mars as debate simmers on whether a meteorite from Mars found in Antarctica harbored microscopic evidence of ancient life and as plans move forward for a series of low-cost, unmanned spacecraft.

Some space experts hold that robots are the best way to explore Mars. But Mr. Goldin and others said initial studies may turn up tasks that only humans can do well. If the possibility of finding Martian life or subsurface water requires deep drilling, it may require the adaptability and intelligence of humans, they said.

"I have a strong feeling there will be things that a robot can't do," Mr. Goldin said.

Some of the technology under consideration for Mars likely will be tested on the International Space Station, construction of which is scheduled to begin in orbit later this year. The 470-ton outpost is to be completed around 2004 by the United States, Russia, the European Space Agency, Canada and Japan.

"The space station is only justified by the idea of a trip to Mars," Dr. Friedman of the Planetary Society said. "Why study the biological effects of long-term space flight on people, and a lot of the other technology, if you are not going to go somewhere?"

One major change being considered for the space station has an important potential Martian application. NASA has issued a stop-work order on a long-planned habitat module being built by the Boeing Company as a major American contribution to the station and is considering substituting a lighter, inflatable unit called TransHab. The pressurized habitat, one of six principal compartments of the station, is to house the living quarters and much of the equipment for American astronauts on the outpost.

NASA officials said they would decide by the end of the year whether to continue work on the original module or to go with the new concept. The cost estimate of either option, to be launched to the station in 2003, is about $100 million, they said.

Instead of having a metal case, TransHab would have a light-weight core made of composite materials surrounded by a shell composed of layers of puncture-resistant fabric, like the cloth used for bullet-proof vests.

If borne out by testing, the technology of TransHab, a term derived from combining the words "transportation" and "habitat," could be used to make living quarters on the Moon or Mars, as well as aboard an interplanetary ship, said engineers at NASA's Johnson Space Center in Houston, where the concept is being developed.

"We're designing a space-inflatable habitat that is safer, cheaper and better than anything currently in the works," said Donna Fender, the project manager at Johnson. "We are not designing Mars hardware, but my goal is to provide an inflatable habitat for the space station that could be used without major redesign to go to Mars."

The module would be compressed around its 11-foot-diameter core when carried to the station by a space shuttle. When released, its inflatable pressure shell would expand into a unit 25 feet in diameter and 26 feet long. The module,

"The space station is only justified by the idea of a trip to Mars," Dr. Friedman of the Planetary Society said.

as tall as a three-story building, would house separate quarters for four to six people.

The TransHab, weighing about five tons empty, would be some 50 percent lighter than the planned Boeing model while providing three times the internal volume. Because of the weight reduction, it can be sent to the station fully equipped instead of being outfitted afterwards, as planned for other station modules, project engineers say. The added space could have multiple uses, from testing large pieces of equipment to recreation, and would provide room for added protection from space radiation, they say.

The design calls for surrounding the central living quarters with a water jacket. Water stops the penetration of radioactive particles, such as cosmic rays and ions from solar flares, and a 4-to-6-inch-thick layer would form a radiation "storm cellar" for astronauts that was not provided for in current plans for the station, Ms. Fender said. Such a radiation shield would be particularly useful on an interplanetary trip or on Mars, which, unlike Earth, has no magnetic field to trap solar radiation above the planet.

Other technology in the works focuses on regenerative life support systems that can recycle wastes while producing oxygen and food for space travelers. NASA engineers working with the Advanced Life Support Program, also headquartered at the Johnson center, say they have made great strides in developing a bioreactor that uses microbes to clean waste water before it is filtered through a conventional reverse-osmosis purification system.

Dr. Don L. Henninger, head of the life-support program at Johnson, said the program completed a 91-day test in December in which four participants in a sealed chamber proved that such a system was feasible. The biological system allowed recovery of 99 percent of potable water, he said, processing a total of 2,300 gallons of water from the 210 gallons originally stowed aboard.

For the first time in such a test, engineers used an incinerator to recycle fecal waste, recovering carbon dioxide and water vapor used to nourish wheat and lettuce growing in an adjacent chamber. The wheat, in turn, produced 25 percent of the oxygen used by the crew.

"We've come so far with this technology that we think we are ready to put some of it on the space station," Dr. Henninger said. "The station can be a testbed for our system while at the same time the system can reduce consumables that have to be taken to the station."

The regenerative system has obvious implications for a Mars trip, because it would be expensive and impractical to take along all of the food, water and other consumable items that the crew would need without recycling it, he said.

The program is planning to build a large research unit known as Bio-plex to test the idea of a completely contained, closed system that can sustain a crew for more than a year. The unit, which would include two modules just for plant production, is being designed to sustain a crew of four with no outside equipment or support, as would be required on the Moon or Mars, Dr. Henninger said.

"Our job is to be ready with the technology when it is needed," he said.

Representative F. James Sensenbrenner, the Wisconsin Republican who is chairman of the House Science Committee, said NASA's current spending on manned Mars work was adequate. "NASA has enough to do now just trying to build the space station and I don't see an interest in starting the wheels into motion for Mars right now," he said.

Advocates of human space flight were disturbed last month by an internal NASA memorandum that appeared to call for an end to studies of future human missions to the Moon and Mars. But Representative Dave Weldon, Republican of Florida, calling such a cut short-sighted, said last week that he received assurances from Mr. Goldin that these studies would continue, heartening proponents of Mars exploration.

"NASA should accelerate the ground work for a Mars mission now," said Dr. Robert M. Zubrin, president of Pioneer Astronautics, a private space research company in Indian Hills, Colorado. "Their current plans are an enormous step forward from a decade ago when they were talking about a $450 billion price tag, but I think it still could be done cheaper and faster. If done correctly, the cost goes down and that increases the possibility of its political acceptance."

III. The International Space Station

Editor's Introduction

When the Soviet Union launched the first component of the permanently manned Mir Space Station in February 1986, Americans watched in disbelief as they were beaten to the finish line once again. In many ways it wasn't the fault of NASA officials: President Ronald Reagan had called for the building of an American space station (which he dubbed Freedom) in his 1984 State of the Union address, but he was soon confronted with serious political ramifications—every lawmaker in the land complained about the station's budget or its design or its purpose. Then, on January 28, 1986, the space shuttle *Challenger* exploded 73 seconds after lift-off, killing all seven crew members. An immediate halt was issued on all human space travel pending an investigation of the most horrific accident in the history of the American space program. In September 1988, NASA launched the first manned American space flight since *Challenger*, but the space station project was in jeopardy. In the aftermath of the *Challenger* disaster, the agency looked for ways to recommit itself to safety—and a risky endeavor like a space station was out of the question.

As the Russians circled the globe in the cramped but efficient Mir, setting human endurance records in space, members of the United States Congress quibbled over the design and cost of an American space station. On several occasions it appeared the project would be killed outright. In 1993 the proposed station came the closest to being canceled when the U.S. House of Representatives passed its funding by only one vote. That same year, the new administration under President Bill Clinton was extremely ambivalent about continuing the Reagan project, particularly as the new president was attempting to bring about closer ties with his country's former cold war enemies. Then NASA officials came up with a brilliant idea: the Russians were working on Mir 2, a replacement to their aging station—why not combine the two programs to cut costs and bolster international relations? Clinton, along with his pro-NASA vice president, Al Gore, supported the endeavor. They not only wanted to develop closer ties with Russia, but also wanted to bring on board a number of American allies who were eager to tie their national space programs to the station. The Clinton administration also supported the measure because they believed that a space station would help employ a vast number of Russian scientists who were put out of work with the collapse of the Soviet Union. (Privately, many in the administration feared that these scientists might build weapons for Iraq and Iran and preferred them to work on something more benign.) As the project began its slow march from the drawing board to the launch pad, it was re-christened the International Space Station (ISS).

Many Russians, on the other hand, were not so happy about this new international project. After all, they had long since been recognized as the pre-eminent power in the development, construction, and manning of space stations and were somewhat unwilling to give up that position. Former cosmonauts and many others in the financially burdened Russian Space Agency grumbled about losing predominance as they transferred from the aging Mir to the International Space Station. Still, the ISS had many supporters in the Russian government, in particular President Boris Yeltsin, who felt that new international cooperation would help bolster the Russian economy and space program.

Section III is intended to familiarize the reader with the history of the International Space Station, its purpose and functions. As the most ambitious international effort ever devised during peacetime, such a project gestures towards a new type of exploration through cooperation instead of competition. Many supporters have argued that the space station is a springboard to the exploration of Mars and other bodies in the solar system; if the springboard is composed of a truly universal infrastructure, then would future missions to other planets come from not one nation, but all of humanity?

"Onward into Space," the first article in this section, was written by Andrew Lawler for *Astronomy*. Lawler gives his readers a solid overview of the political history behind the space station, discusses potential benefits and costs, and briefly outlines how and when it will be constructed. The next two articles were published in *Popular Science*. In the first, "The Sum of Its Parts," Jim Schefter goes to great lengths to explain how the ISS will be assembled, which countries are building what components, and what the general operations will be when the first crews begin occupying the station. In the second *Popular Science* article, "Why Build a Space Station?," William E. Burrows touts the merits and importance of this project. "Like the *Apollo* program that put a man on the Moon," Burrows notes, "the International Space Station is a massive engineering project that will keep a handful of people in space and many thousands employed on Earth. In this post–cold war era, when the U.S. aerospace industry is in decline, the space station program has created thousands of jobs for highly skilled workers." On the other hand, Gregg Easterbrook, in "Cosmic Clunker," his editorial for the *New Republic*, believes that a space station of this size and magnitude is riddled with hidden costs and would in fact absorb more money than it gives back. He noted that "the station's sticker price is $21 billion, but $40 billion is more realistic when launch costs are taken into account." He goes on to mention that every member of the crew needs several gallons of bottled water a day and concluded the "estimated annual bottled water cost for the station: $817 million. Is this a defensible use of public funds?" This section concludes on a more optimistic note, with John M. Logsdon of *Aviation Week and Space Technology* writing in favor of the space station in "Building a Space Station Still Makes Sense." In this editorial, he notes that the space station is simply the next logical step, if humanity is to continue going into space.

Whatever one's point of view, it appears that the International Space Station will be a source of international debate well beyond its projected completion in 2004, but it cannot be disregarded. It is arguably the single largest engineering feat of the 20th century and truly one of the most ambitious project of the space age. How history will judge it remains to be seen.

Onward Into Space[1]

Forget the elegant wheel-shaped craft from *2001: A Space Odyssey*. The International Space Station will look like a flying trailer park when space walking astronauts finish assembling it. Awkward metal trusses will connect utilitarian, pressurized, barrel-like modules to ungainly, winged solar panels that will convert sunlight into electricity.

Six-astronaut crews will have priceless views 200 miles above Earth aboard this ultimate mobile home. Critics may attack it as pork-barrel space junk, but the space station will be the single largest international scientific and technological effort ever undertaken in peacetime. However, the drumbeat of criticism is unlikely to fade even if astronauts finish it on schedule in 2004. This timetable, while optimistic, means the station will be a decade late and cost tens of billions of dollars more than originally conceived. Total cost estimates range from $40 billion to more than $100 billion. However, NASA managers proudly assert that it will mark a milestone in the human push to explore and live beyond the planet.

Space-station partners hope to capture the world's imagination with the first of what will be dozens of assembly and maintenance launches. It is more than a testing ground for future missions to Mars; It is a scientific United Nations and a milestone of scientific, technological, and political ingenuity of the highest order. That is if the partners can pull it off. Unfortunately, Russia can't deliver the Service Module, its primary contribution, in time for its planned April 1999 launch. NASA wants Congress to give Moscow a $660 million bailout over five years. The partners are Japan, Canada, Belgium, Denmark, France, Germany, Italy, the Netherlands, Norway, Spain, Sweden, Switzerland, and the United Kingdom.

Voodoo Science

President Ronald Reagan championed the space-station idea in his 1984 State of the Union address. The critics say the project's 15-year history is a testament to political creativity, not scientific achievement. "Its goals are political, not scientific," says Robert Park, a longtime station opponent who works for the American Physical Society.

---

1. Article by Andrew Lawler. From *Astronomy* p42-51 Dec. 98. Reproduced by permission. Copyright © 1998 *Astronomy* magazine, Kalmbach Publishing Company. Andrew Lawler is a fellow at the Knight Science Journalism Program at the Massachusetts Institute of Technology. Lawler also writes about science policy for *Science* magazine.

"It's not a science project— it's a foreign policy initia- tive," says John Pike, an analyst with the Federa- tion of Amer- ican Scientists. "For Reagan it was to whip the Evil Empire; for Clinton it is to be friends with the Russians."

Park and many other scientists wish the space-station money was being spent on productive, inexpensive planetary science missions. They complain that there is little evidence that the station will benefit life and materials sciences. Park rebuffs claims made by lawmakers during funding feuds that the program will help find cures for diseases and result in exotic new materials. "It's simply not good to promise things that will not happen," he says. "This is clearly voodoo science."

Nobody disputes that research is not the real reason for the space station; otherwise it never would have survived, says a coterie of Washington observers. "It's not a science project—it's a foreign policy initiative," says John Pike, an analyst with the Federation of American Scientists who has watched the program closely. "For Reagan it was to whip the Evil Empire; for Clinton it is to be friends with the Russians." It also has provided thousands of jobs to American aerospace workers from California to Florida, and those jobs amount to political muscle.

In the early 1980s, aerospace money began pouring into the Pentagon and Southern California's aerospace industry. NASA's new administrator at the time, James Beggs, saw an opportunity for his agency to win approval for an expensive new program that would give the freshly minted space shuttles something to do. Beggs knew NASA needed to sell the idea and he knew the potential buyers. Scientists wanted a research platform in the weightlessness of space. Eager aerospace contractors wanted federal money. Congressional incumbents, facing re-election, wanted research and development dollars for their states and districts. The State Department wanted a way to strengthen United States' strategic alliances and counter the Soviet threat. Allies of the United States wanted to join the exclusive club of nations able to send humans to space. Engineers wanted to play in the ultimate playground, and space enthusiasts were eager for a new adventure.

Beggs methodically gathered his allies under the space-station umbrella. This was not to be an impulsive dash like the Apollo missions to the Moon in the 1960s and '70s. This was not to be the mere half a loaf that President Richard Nixon offered NASA in the 1970s after he rejected the space station and approved a scaled back shuttle. Instead, the station was a carefully calculated effort to win political backing for an expensive project that garnered little enthusiasm elsewhere in the government.

It almost failed. At a heated White House Cabinet meeting on a cold winter day in 1983, Beggs found himself under attack by Caspar Weinberger. The defense secretary feared the project would cost far more than the $8 billion Beggs had estimated. Weinberger worried it would eat into defense spending. At some point during the meeting, Reagan winked at the NASA administrator, a sign that he should not worry about the carpers.

A few weeks later on national television, in words reminiscent of President John Kennedy's 1961 call for a Moon landing, Reagan urged the nation to build the space station in one decade. It was to be America's high-technology answer to the Russian space station, which by then had been in orbit for several years. The space station also was a way to cooperate with allies of the United States that were nervous about the hawkish Reagan. "It was the cold war that kept it going in the 1980s," says Joe Rothenberg, NASA's new space flight chief.

Space-station dollars flowed into 42 states. "In the 1980s, jobs were the primary driver," says Bill Smith, a former Democratic staff in the House of Representatives. Most of those jobs were in politically powerful areas in California, Florida, and Texas. Lobbyists from McDonnell Douglas, Rockwell, General Electric, and Grumman cruised the halls of Congress, keeping tabs on the "economic impact" of the program. That pressure helped supporters overcome yearly attempts by critics to cancel the project in Congress.

Tough Sledding

By the turn of the decade, the United States already had spent nearly $8 billion on the project. (Reagan dubbed it Freedom to underscore its challenge to the Soviet political system.) However, the *Challenger* explosion in 1986 and the resulting fiery death of the five man, two-woman crew prompted upheavals in NASA's management. Members of Congress quibbled over the station's design, and contractor squabbles slowed the project. As a result, costs soared. "It became, as one NASA manager put it at the time, a self-eating watermelon," recalls Adam Gruen, a historian paid by NASA to write the inside story of the program. "When it takes you 12 months to delay the schedule by 10 months, that's tough sledding."

The second crisis was the breakup of the Soviet Union. It came as American budget deficits pinched domestic spending and squeezed NASA's budget. At the same time, costs to complete the space station had doubled to nearly $14 billion since the *Challenger* disaster. In 1993, with the Soviet threat gone, the space station coalition began to unravel. Bill Clinton took up residence in the White House amid talk of massive cost overruns. The fat, expensive, slow-moving space

Reagan urged the nation to build the space station in one decade. It was to be America's high-technology answer to the Russian space station, which by then had been in orbit for several years.

station was an easy target for budget cutters. The new president and his science advisers appeared to favor faster-paced efforts that could more quickly benefit United States competitiveness. "The president needed to know why the space station costs so much," one former White House official recalls. "He was told it needed a brain transplant."

With Clinton's election, the faces around the Cabinet table had changed. Instead of Beggs sparring with Weinberger, Vice President and technophile Al Gore squared off against White House Budget Director Leon Panetta, a longtime station opponent while he was a member of Congress. Panetta argued in a February 1993 Cabinet meeting that Clinton should cut the space station's budget by half. Gore knew such a blow would ruin the program's carefully built constituency. Gore and Panetta agreed to a compromise—NASA would cap space-station expenditures at $2.1 billion a year, redesign the project, and present the new option that summer to Clinton.

Clinton's advisors met again in June to look over the new plan. Clinton talked about the need for science and technology investment. However, in private sessions prior to the gathering, his advisers say he worried about the ugly political fallout of canceling the station. By that time, more than 15,000 American workers in 42 states depended on the program for their livelihoods. A key space-station supporter, Texas Governor Ann Richards, visited Clinton the day before the space-station meeting at the White House. In countless phone calls during the preceding months, Richards had badgered the president. She desperately wanted him to spare the project, White House aides recall.

There was more than jobs at stake. Clinton's NASA Administrator, Daniel Goldin, urged United States and Russian officials to discuss the merger of Russia's Mir 2 with the space station. The appealing new objective catapulted the space station beyond mere jobs and science. Clinton gave the green light to revamp Reagan's space station and proceed.

A Black Hole

That decision was nearly a Pyrrhic victory for program supporters. Armed with the cost overruns and emboldened by months of White House ambivalence, congressional critics attacked a few days later. The program's bête noire, Indiana Democratic Representative Tim Roemer, compared the program to "a black hole quickly sucking away money from other promising scientific projects." He argued that NASA had mismanaged the effort and that there was no sane reason to put humans into orbit. House lawmakers came within a single vote of canceling the program.

"[President Clinton] needed to know why the space station costs so much," one former White House official recalls. "He was told it needed a brain transplant."

NASA and White House officials realized that jobs and rhetoric about technological leadership were no longer enough to save the program. Goldin found a new purpose for it. He understood the powerful symbolism of joining two old enemies in a high-tech project. There also was growing concern that thousands of poverty-stricken Russian scientists and engineers might design missiles and bombs for Iran or Iraq. The United States and Russia, Goldin argued, were already cooperating in their shuttle and Mir programs—why not extend that venture to the space station?

It worked. White House officials embraced the idea that the space station could be a strategic corner in national-security policy. The annual budget cap of $2.1 billion defused criticisms of its accelerating cost. NASA set up a permanent "war room" on the seventh floor of its Washington headquarters building, a short walk from the Capitol, to manage the fight to save the station. Roemer and fellow longtime opponent, Democratic Senator Dale Bumpers of Arkansas, lost the upper hand.

Cost Savings Vanish

However, promises of cost savings never materialized. Delays again set in, and Russian participation became a financial burden. This spring, an independent audit revealed that the program would cost as much as $7 billion more than NASA estimates and take up to three more years to complete. Boeing, which took over in the early 1990s as the prime contractor, has had difficulty controlling the costs of the hundreds of suppliers and subcontractors. Meanwhile, the $2 billion that Goldin estimated would be saved with Russia as a partner vanished.

"We must get this program under control," thundered Jim Sensenbrenner, the Wisconsin Republican who chairs the House Science Committee, in a recent hearing. Roemer cited the report as proof that the space station was beset by budgetary overruns and massive scheduling problems. "The more daylight that shines on this program, the darker the outlook for its successful completion," he said.

Somehow, Sensenbrenner's and Roemer's criticisms seem to bounce off NASA officials, who know that the president and most lawmakers support the program. The program, says former White House Office of Management and Budget chief Franklin Raines, "is bound to face technical, cost, and scheduling challenges." White House officials say privately they don't know how they will come up with the extra money to fund the space station. Researchers who depend on NASA funding worry that the Clinton administration eventually will cut science projects to pay for the overruns.

Station Assembly Schedule

Some of the major components of the International Space Station are numbered in this artist's concept as an assembly and flight-date guide for individual modules . . . The dates of the flights were set by NASA.

1) FLIGHT 1R—1999 The Service Module is a living quarters and the primary Russian contribution to the station. It contains a propulsion system to move and properly orient the orbiting station, and it is the primary docking port for resupply vehicles.

2) FLIGHT 4A—1999 The Integrated Truss Structure will provide cooling radiators, a communications system, and power with its solar arrays and batteries.

3) FLIGHT 8A—2000 Astronauts will attach the central-truss segment to the United States' laboratory. The truss segment will be the kingpin of the 300-foot Integrated Truss Structure. Astronauts will attach four additional starboard truss segments and three port segments later.

4) FLIGHT 1J—2002 The primary Japanese contribution to the station is the Japanese Experiment Module. Its robotic arm will tend experiments on the laboratory's adjacent "back porch," or Exposed Facility.

Already, NASA has postponed construction of key scientific facilities aboard the station in order to pay for building hardware. The agency's other space-science efforts, like robotic missions to Mars and a program to monitor Earth's environment, are in danger, Park and others warn.

The Right Stuff

NASA's space-station chief Randy Brinkley, Rothenberg, and other NASA officials say the threat to scientific research would be greater without the space station. They argue that the more money NASA has received for human space missions, the more it also has received for science overall. However, critics like Park dispute that argument. The *Pathfinder* mission to Mars last year captured the public's attention for weeks, and the bread-box-sized *Sojourner* rover became a media star. Newspapers published obituaries in observance of *Sojourner*'s passing. "NASA hopes the space station will recapture the imagination of Americans and bring back the days when we crowded around our living room TVs to watch Neil Armstrong," says Roemer. But it was *Pathfinder*, not a human mission, that drew crowds July 4, 1997, when Pathfinder landed, he adds. "I believe the American people would support the space program without people in it," Park insists.

5) FLIGHT 20A - 2002 Node 3 will contain avionics and life-support equipment. The node will be the attachment point for the United States Habitation Module, the crew-return vehicle, and other additions.

6) FLIGHT 1E - 2003 The Columbus Orbital Facility will provide additional research capability. It is the European Space Agency's primary contribution to the station.

7) FLIGHT UF-7 - 2003 The Centrifuge Accommodation Module attaches to Node 2 and will complete the station's laboratory facilities. The module will control gravity for research experiments.

8) 1999 - 2003 At many points during the assembly process, astronauts will add photovoltaic panels that convert sunlight into electricity.

9) 1999 - 2003 Heat-dissipating radiators will be added at various points in the assembly process.

Six astronauts endlessly orbiting Earth may lack the valor of a Moon-walking Buzz Aldrin, but human space flight provides an element of danger, drama, and daring lacking in robotic missions.

NASA has no intention of testing that theory. It built its reputation on heroic astronauts with the right stuff. Taking joysticks from heroes and giving them to robots is like asking the Air Force to replace its top-gun pilots with automated drones. Politicians also are unlikely to back that approach. Even Goldin, who has pushed for faster, better, and cheaper space missions, is eager to exempt the over-budget space station because he regards it as central to the agency's future.

"It's the human face of the space program that has the most appeal," says Smith. Six astronauts endlessly orbiting Earth may lack the valor of a Moon-walking Buzz Aldrin, but human space flight provides an element of danger, drama, and daring lacking in robotic missions. Launches involving robotic missions are sparsely attended while space shuttle launches routinely draw huge crowds.

"It is the 20th century version of the pyramids of Giza or the cathedrals of Europe," says historian Gruen. "It's [a] technological program employing tens of thousands of skilled laborers, and a bold cultural statement of vision and aspiration." He adds with a laugh, "With enough time and money, you can do anything."

Assembling the Puzzle

The International Space Station is not like a snap-together Lego or Tinker Toy set. It is a complicated, 1 million-pound jigsaw puzzle. It will consist of tubes, girders, cables, and delicate, wing-like solar panels. Workers in Moscow, Los Angeles, Turin, Nagoya, and a dozen other cities are making most of the pieces. For the first time, astronauts will connect many of the parts in space without benefit of prior ground trials. Outfitted in bulky space suits, wearing fat, cumbersome gloves, they will fasten, weld, and tie the components together as they space walk above Earth.

United States' space shuttles, European Ariane rockets, Japanese launchers, and Russian Soyuz vehicles will relay astronauts and deliver the space-station components in more than 50 flights. An additional 30 flights will provide fuel, food, and other supplies during construction. "Think of the sheer magnitude of it," says Joe Rothenberg, NASA's soft-spoken and bespectacled space flight chief. "There are some 60 major pieces and it can't be assembled first on the ground and tested."

It will not be easy. Rothenberg oversaw the dramatic Hubble Space Telescope repair mission that corrected the effects of that telescope's improperly made and inadequately tested primary mirror. "On the Hubble mission we pretty much knew what to expect, but with the space station, astronauts will have to be able to fix a variety of problems on the spot," says Rothenberg.

The first United States–made pieces began dribbling into the Kennedy Space Station Processing Facility in Florida this past spring in preparation for launch. The first step in the assembly plan is to send up the Russian-made, 20-ton module that contains electronic equipment, fuel, and control systems. Next, a shuttle will deliver a small round node that will serve as a construction shack. Astronauts will attach larger modules to it. Later, they will attach a truss and solar panels to juice up the nascent station's electrical systems.

At this point, astronauts will attach the service module—a large structure designed to house the early station's control and support equipment. This module is the centerpiece of Russia's contribution. The Russian government's chronic lack of money led to delays in the completion of the module, but NASA officials hope it will be ready for launch by the summer of 1999.

Astronauts will attach the United States' laboratory, the centerpiece of the American program, after they add more solar panels. Next, crew members will add United States and Canadian robotic arms, airlocks, and eventually the United

> *"Think of the sheer magnitude of it," says Joe Rothenberg, NASA's soft-spoken and bespectacled space flight chief. "There are some 60 major pieces and it can't be assembled first on the ground and tested."*

States' habitation module. Early next century they will attach a lab built by the European Space Agency and another built by Japan, completing the station. The assembly will take at least five years, barring a disaster. (A launch failure could set the program back months or years.)

Compared to the cramped and aging Russian Space Station, Mir, the new space station will be a luxurious campus. Astronauts will gather for meals around a table, sleep in compact but private crew quarters, shower, and occasionally "walk" outside. They will work in half a dozen other modules conducting experiments and performing maintenance.

The United States will own half the real estate, including the laboratory module, the sleeping quarters, and three smaller cylinders called nodes that will connect the modules. NASA also will own a 20-ton control module built by a private Russian company.

The Russian government is providing two research modules, a critical service module with life support systems, and transport vehicles to carry supplies and dispose of trash. Japan and Europe are building their own scientific modules, along with transport vehicles. But given the political and financial turmoil in Russia, there are doubts that the world's first spacefaring nation will be able to fulfill its promises.

Pitching the Station

"You're looking at the future," says the NASA guide to dozens of visitors dressed in shorts and T-shirts. I squint through glass and look downward. In this vast building at Kennedy Space Center in Florida, powder-blue-suited men and women walk like ants on the spotless floor among pieces of the International Space Station. Ironically, most of these visitors are here because of a glitch: NASA scrubbed today's shuttle launch. This excursion transforms disappointment into awe.

In 2004, astronauts are scheduled to finish connecting a gangly collection of tank-like pods, metal frames, and solar panels. The result will occupy the volume of a major-league baseball park. Our guide pitches the future benefits of the space station and tells us to anticipate the first launch in June 1998. (That launch date, doubtful at the time, later was delayed for months.) He then asks us to leave through a side door so another group of visitors can enter.

I gawk a moment longer from our box-seat, plate-glass vantage point high above workers assembling pieces of white hardware. I ask the guide if he is aware that the first launch almost certainly will not happen before late fall. He shrugs. The future as described by NASA sometimes takes longer than promised to arrive.

Biggest, Best, Costliest

It is difficult to be straight-faced and unimpressed. The 457,000-square-foot, white-painted Space Station Processing Facility hangar, trimmed in red steel beams, soars 10 stories above the swaying marsh grasses and scrubby mangroves of Florida's coastal wetlands. The Kennedy Space Center building bristles with metal cranes and national pride. As workers piece together the many parts of the soon-to-be-orbiting International Space Station, visitors jostle in an adjacent building to see a mockup of a training lab used by Apollo astronauts.

The space station's stated goal is to improve humanity's quality of life through international research and good will. Here, in this publicly funded oasis of technology, the future appears to be well in hand.

The influence of Disney World in nearby Orlando, Florida, also seems to have spilled over to Kennedy where workers are painstakingly assembling NASA's Tomorrow Land. However, this will be no amusement-park ride. This will be the largest, most expensive, and most complicated international scientific and engineering project ever built. The project already has consumed 15 years and billions of dollars, rubles, marks, and yen. It somehow has endured changes of government, currency fluctuations, and funding crises from Moscow to Paris. Key members of the Clinton administration once tried to cancel it and members of Congress bitterly attacked it—unsuccessfully.

The space station's improbable ride of political twists, technical leaps, and diplomatic dips surpasses the loops and curves of the rides at Disney World's Space Mountain. The space station's appeal is so multifarious it can contain conspicuous contradictions. It is a Cold War relic that depends on Russia. It is a scientific mega-project dismissed by many researchers as an utter waste of money. Even though NASA has otherwise dedicated itself to faster, better, and cheaper missions, the survival of the bloated, over-budget, and behind-schedule space station suggests that biggest is best.

The project fulfills our desire to establish the first international colony in space, but fatalistic critics call it an unstoppable political steamroller. "We are not going to kill the space station," a longtime opponent, Democratic Senator Dale Bumpers of Arkansas, told NASA Administrator Daniel Goldin at a recent Senate hearing. "You have nothing to fear."

The space station's stated goal is to improve humanity's quality of life through international research and good will. Here, in this publicly funded oasis of technology, the future appears to be well in hand.

What Will the Space Station Do?

"It's a means to an end," says Randy Brinkley, leader of the space station program based at Johnson Space Center outside Houston. "It's a means to a new way of exploration and cooperation, a place to learn about the space environment, conduct space science, observe planet Earth, and use the microgravity environment."

Brinkley envisions a beehive of scientific activity conducted by Americans, Russians, Germans, French, Japanese, Canadians, and a half-dozen other nationalities. They will study the effects of near-zero gravity on baby rats and grow crystals that could lead to better drugs. Outside, cameras aimed at Earth and the stars will maintain a watch.

The crew members themselves will be guinea pigs. Scientists will study the effects of radiation and microgravity on them. Such research, NASA researchers enthusiastically say, could provide data necessary to plan lengthy expeditions to Mars. They add that the information also could lead to the creation of new products, and provide clues to curing diseases like osteoporosis, diabetes, cancer, AIDS, parasitic infections, and emphysema.

Such optimistic hopes irk many researchers. They doubt the efficacy of space-based research. Nevertheless, breathless NASA press releases predict that the medical research will help ease earthly afflictions while new materials and processes developed on the space station will "create jobs and economic opportunities worldwide."

The Sum of Its Parts[2]

Late June in Kazakhstan. A Russian Proton rocket sits poised for launch at the once-secret Baikonur space base. Observers from NASA and other space agencies have crowded into the vintage control room, and journalists from around the world are monitoring the countdown. At zero, the Proton's engines thunder; it lifts away from Earth and quickly disappears from sight. The first module of the long-awaited International Space Station is headed for orbit.

If all goes well, that will be the scene just a month from now, when one of the world's most ambitious engineering projects finally gets off the ground. In January 1999, after six launches by U.S. space shuttles and Russian boosters, a three-man crew will take up residence. From that moment onward, the space station will be a permanent off-planet extension of human civilization.

When the station is completed, around the end of 2003 if NASA sticks to its schedule, it will be a multiroom hotel and research facility orbiting the Earth every 90 minutes.

When the station is completed, around the end of 2003 if NASA sticks to its schedule, it will be a multiroom hotel and research facility orbiting the Earth every 90 minutes. Its permanent population will be six or seven, with the mix shifting as Americans, Russians, Europeans, and Japanese move in and out over the months and years. By that time, resupply and assembly flights by shuttles or Russian rockets will have become so routine that today's breathtaking series of space rendezvous is certain to become tomorrow's humdrum.

You'll know the ISS is there. Five times the size of the Russian space station Mir, it will have solar wings and radiators sprouting from trusses spanning 356 feet. Its central core will be a collection of motor-home-size labs, living quarters, and supply canisters plugged together like a supersize Lego set. Together they'll fill a fore-to-aft length of 290 feet, almost as long as a football field.

The final configuration, requiring 45 separate space launches over at least six years and many long days of spacewalks by astronaut construction workers, will track diamond-bright across the night sky. Only the Moon and Venus will be bigger and more visible.

"It takes a while to get perspective on how big this thing is," says Kevin Chilton, the station's operations manager at the Johnson Space Center in Houston. To demonstrate, he stretches his arms across a model of the station without reaching either end. The space shuttle docked to the station

---

2. Article by Jim Schefter. From *Popular Science* p57-61 May 1998. Copyright © 1998 L. A. Times Syndicate. Reprinted with permission.

is only the size of a football and almost lost in the labyrinth of trusses, panels, and modules.

The station's complexity is as awesome as its size. Built by a partnership of 16 nations, ISS will consist of hundreds of individual elements that come from all over the world. "Most of the elements will never be physically mated until they come together in orbit," explains Randy Brinkley, NASA's Houston-based program manager for the space station. "But we're confident that when we turn on the lights, it will all work."

NASA is relying on Boeing, its prime contractor for the station, to ensure that all of the pieces fit together properly and work exactly as planned. Boeing has plenty of experience building complicated flying machines from parts made by far-flung suppliers: The company's 777 airliner, introduced four years ago, was designed and digitally preassembled on computers. "We're borrowing heavily from our brethren in Seattle on the 777 program," says Douglas Stone, Boeing's program manager for the space station.

But even if all the pieces fit perfectly, the assembly process itself will be risky. Unlike an airplane, the space station is flown while it's being built, and each new piece that is added may change the way the station behaves in flight.

Computer software problems are also a concern, "probably the biggest single risk that we face on this project," says Stone. Like the hardware elements, the software components come all over the world and must mesh smoothly. The astronauts won't even be able to turn on the station's lights without help from a computer, because all of the electric power switching will be controlled by software.

Working hand in hand with other countries will be another challenge. "This station is really a new era for all of us," says Chilton. "We've had to learn how to be good guests on Mir and at the Russian training and command centers. Now we're going to be equal partners."

The station is divided roughly into two sections, one built primarily by the United States and the other built by Russia. "In some ways, it looks like two different spacecraft in close proximity," says Stone, "but to the crew, it will all be one vehicle."

The other nations participating in the project are also contributing hardware: Japan's space agency, NASDA, is supplying a laboratory module that includes a "back porch" where experiments can be exposed to space; astronauts will use a robotic arm to lift experiments onto the porch from an air lock. The European Space Agency is also providing a lab module. The Canadians are building a robotic arm and hand

The station's complexity is as awesome as its size. Built by a partnership of 16 nations, ISS will consist of hundreds of individual elements that come from all over the world.

that will be invaluable for assembly work and heavy lifting. And Brazil is contributing a window facility; astronauts can use it to study changes in the Amazonian rainforest, among other things.

The first building block to be launched will be a U.S.-funded, Russian-built pressurized module called the Functional Cargo Block (or FGB, from the Russian name). Built under subcontract to Boeing by Krunichev Industries near Moscow, the FGB will be a free-flying satellite until the second piece of the station arrives.

About the size of a long camper trailer, the FGB is a 20-ton multipurpose power and propulsion plant that gets the station assembly work started. A pair of solar panels will unfold in orbit to provide electrical power. The FGB carries propulsion and fuel storage—to keep the early station components in orbit—and docking ports at each end of its canister shape.

In July, the shuttle *Endeavour* will haul a module called Node 1 into orbit and dock it to the FGB. U.S. astronauts Jerry Ross and Jim Newman will spacewalk to the fledgling space station and connect electrical cables between the two units. Their initial foray into space construction will come late in the shuttle mission.

"We've learned a lot from the Mir experiences," says the station program's medical officer, Dr. John Charles, who monitored U.S. astronauts assigned to long-term duty on the aging Russian station. "For instance, it takes the average person about four days to make the bodily adjustments to space. So we don't do spacewalks in the first four days." That rule will be followed throughout space station construction.

Node 1, built by Boeing in Huntsville, Alabama, is a pressurized module with six ports. One mates with the FGB, and the opposite port will be the early connection between shuttles and the station. Other ports will eventually get docking units to which more modules will be attached.

As the assembly continues, the node will sprout the long main truss holding the solar arrays and thermal radiators; a U.S. science lab; a habitation module where the astronauts eat and sleep; and other modules. A second node will be brought up in 2001 as the station grows.

"It's not the way you'd build your house," says Brinkley. Most housebuilders would start by erecting an outer shell; finishing the individual rooms would come much later in the construction process. But to construct a space station, says Brinkley, "you build the bathroom, and live there at first." Then you add the other rooms, one module at a time.

When *Endeavour* leaves the mated FGB and the first node behind, the longest quiet period in the assembly process begins. It will be at least several months before the third station element, the Service Module, is sent up from Baikonur. Using automated docking procedures perfected by the Russians, the Service Module will mate to the FGB port opposite from Node 1.

The Service Module, which is based on technology used in the Mir station's core module, carries the environmental controls and life support system for the entire station. It also has the primary docking port for Russia's *Soyuz* spacecraft, and the fuel and rockets for both attitude control and station reboost. Reboost is needed periodically to raise the space station's orbit and keep it from reentering Earth's atmosphere.

The Service Module has been the most problematic component of the station, because the Russians have fallen behind schedule in building it. Last year, NASA became so concerned about the delivery of the Service Module that the agency negotiated a $170 million contract with the Navy to develop an alternative—by modifying a Navy spacecraft built in the early 1980s.

"It was a spinning upper stage to boost shuttle-launched Navy satellites into higher orbits," says Ed Senasack of the Naval Research Laboratory. The spacecraft was abandoned after the Pentagon canceled military use of the shuttle in the wake of the *Challenger* disaster, and was headed for the scrap heap. "It needed some modifications, but we saw immediately that it could fill in for the Russian Service Module," Senasack recalls.

For now, the Service Module seems to be on track. But the modified Navy hardware, renamed the Interim Control Module, is still under development. If the Russians falter again, the ICM will be ready for flight. If not, the ICM can provide additional storage and boosting capabilities for the space station.

Assuming the Service Module is delivered in time to be launched in December or early 1999, it will be docked to the first two pieces of the station. Its solar wings will then be deployed, doubling the electrical power capacity of the connected units.

Then the business of building a space station gets hectic. A total of 14 shuttle and Soyuz booster missions to the ISS are scheduled for 1999. In January, the shuttle *Atlantis* will haul up a variety of components to be plugged into Node 1's ports or to be installed inside. Among these are the first truss structure and a third docking port.

The docking port will later be used by a shuttle to offload the first large solar array provided by the United States. It will be attached to the truss in April 1999. Made by Lock-

The Service Module has been the most problematic component of the station, because the Russians have fallen behind schedule in building it.

heed Martin Missiles & Space, the solar panels are roughly 100 feet long, the largest ever built for space. A total of 16 panels will eventually be installed, covering about an acre.

While the U.S. solar arrays will provide the electrical power for life support and scientific experiments aboard the station, Russia will supply the propellant fuel to keep the structure in orbit. "We're the power company, and they're the gas company," explains Brinkley.

But long before all of the solar panels are installed, the ISS will open for business. "The crew will come up immediately," manning the station as soon as *Atlantis* delivers its load of critical hardware in early 1999, Brinkley says. Barring any major emergencies, the ISS will never again be without human occupants.

The station's first commander will be an American, Navy Captain and former SEAL Bill Shepherd. He and two Russians will ride into orbit aboard a *Soyuz* spacecraft, which will be docked to the station and serve as an emergency rescue vehicle. Future crews may ride up in a shuttle and come back in a *Soyuz*, or vice versa.

When the U.S. Laboratory Module is attached in mid-1999, the station will begin looking something like its artists' portraits. The first crew will leave a month later, and a team of one Russian and two Americans will take over, probably for a five-month stay. "We're still staying with our basic agreement with our Russian partners," says Chilton, "that during the assembly phase, we will share time on orbit fifty-fifty for the crews."

A U.S.-built crew-rescue vehicle will be added to the station in 2003. After that, a six-person crew will occupy the station. Some astronauts may stay as long as 187 days, but there are no plans yet for longer missions. (The official life expectancy of the station itself is 10 years, but it should last much longer.)

"The Russians say that six months is the maximum-efficiency stay in space," Dr. Charles says, "but I think it could really be years. The key is exercise, and there's probably no time limit for a healthy person up there."

Dr. Al Holland oversees crew psychology at the Johnson Space Center. Working with the Russians and with other nations is bound to create some problems, he thinks. "The big challenges are going to be with the first few crews," he says, "and those will be organizational—setting up shop, solving hardware problems, talking to control centers in Moscow and Houston."

After that, life aboard the ISS should settle into a routine. Holland likens it to a naval deployment. "You're not going up there to drive something," he says. "You're going up to do a job. In the end, a space station is like a ship, not like a shuttle."

But these space sailors are building their own ship from scratch, in the middle of an unforgiving empty ocean. Getting from here to there is going to be a challenge.

Why Build a Space Station?[3]

This massive project isn't just for science and space exploration. It's also about jobs and international stability.

Ever since *Sputnik*, and especially during the cold war, the score in space has been kept by setting records. To all of the others—Yuri Gagarin's epic flight, *Eagle*'s landing on the Moon, *Voyager 2*'s grand tour of the outer planets, and dozens more—add the International Space Station: When completed, it will be the biggest and most complex structure that has ever been built in space.

When completed, it will be the biggest and most complex structure that has ever been built in space.

But the station holds another record that is anything but enviable: It has caused the longest and most rancorous battle in the history of the space age. Although no space project is wholly divorced from politics, the space station has been at the heart of a particularly drawn-out conflict ever since President Ronald Reagan endorsed its construction in December 1983. Only in the final hour, now that the project has moved off the drawing board and onto the launch pad, does the station enjoy the support not only of the White House but also of Congress, a majority of the American public, and 15 other nations. That's a reflection of how the station's mission and potential have grown over time.

Even the station's name reflects the fact that its 15-year gestation period, through one design change after another, spans two dramatically different times. During Reagan's era, the station was named Freedom as an in-your-face gesture to the Soviet Union. But now, with the cold war over and the United States eager to spread the escalating cost of the program to other countries, including Russia, the project has been rechristened the International Space Station as a gesture toward reconciliation.

As the station's name has changed, so has its purpose. The program is no longer solely about conducting scientific research in the absence of gravity, or about learning how to build a massive structure in orbit, or about establishing the base for a future trip to Mars. The International Space Station is more than merely the next great adventure of the space age. It's also about promoting international cooperation and finding peacetime work for former enemies.

Like the Apollo program that put a man on the Moon, the International Space Station is a massive engineering project that will keep a handful of people in space and many thou-

---

3. Article by William E. Burrows. From *Popular Science* p 65-69 May 1998. Copyright © 1998 L. A. Times Syndicate. Reprinted with permission.

sands employed on Earth. In this post-cold war era, when the U.S. aerospace industry is in decline, the space station program has created thousands of jobs for highly skilled workers.

"Not a single dime of this money is spent in space; it's all spent right down here on Earth," says Saunders B. Kramer, a retired research scientist who designed and patented a space station in 1958 while working for Lockheed Missiles & Space. "The guys get their salaries and they go to the tailor, the butcher, the candlestick maker, and the clothiers and the furniture maker and everybody else."

The program has also created jobs in Russia for engineers who might otherwise be recruited to work on foreign military programs. In fact, many of the U.S. and Russian engineers who are now working side by side on the space station are former defense workers who once targeted each other. The station engages 16 of the nation's most technologically advanced nations—the United States, Russia, Canada, Japan, Brazil, and 11 European nations—in a peacetime project of massive proportion.

"It's a political program as well as a science program," says Douglas C. Stone, program manager for the space station at Boeing, the company supervising the station's construction as NASA's prime contractor. "Just getting these countries to work together peacefully is worth more than the price of the station," says Stone. In fact, it amounts to unprecedented international cooperation: the first step toward what space visionaries call a unified "planetary civilization" that will explore space as citizens of Earth, not of individual nations.

To NASA, the space station is a step that follows logically upon the heels of the space shuttle program. Space station supporters correctly point out that one of the space shuttle's key missions is to truck astronauts and hardware into orbit, and that its usefulness (and huge taxpayer investment) will be greatly diminished unless a station is built.

Even so, supporters have at times had difficulty justifying the cost of the station—which will probably mount to more than $40 billion before it is completed. As it is, the station's price tag has climbed over the years from an initial estimate of $8 billion to design and develop it (but not construct it in orbit) to the current U.S. contribution of about $17.4 billion (the figure fluctuates, with Boeing already racking up $600 million in cost overruns). The Russian share is roughly $10 billion, while the European Space Agency is contributing $3.77 billion; Japan $3.1 billion; Canada $850 million; and Italy $550 million separately, even though it is a member of the European group. Congress has earmarked another $13

In this post-cold war era, when the U.S. aerospace industry is in decline, the space station program has created thousands of jobs for highly skilled workers.

billion, or $1.3 billion a year, for operating the station through 2012.

It is Congress that Charles D. Walker, a payload specialist on three shuttle missions and a Boeing senior executive for advanced business development, blames for the station's spiraling cost. "They [Congressmen] have no experience with technical programs, no experience with government and industry working on technical design, development, and deployment," yet they want to micromanage the huge program, he says.

But there is plenty of blame to go around. One White House budget official, Frederick N. Khedouri, complains that NASA consistently underestimates the program's cost. "They always talk about a stripped-down model that doesn't exist," he says.

Scientists opposed the station because they felt its huge expense would drain money from much more useful research programs.

The debate has been going on at least since April 1960, when representatives from the fledgling space industry held a manned space station symposium in Los Angeles sponsored by NASA, the Rand Corp., and the Institute for Aeronautical Sciences. Squabbling broke out regarding what a station should be, where it should be put, and how to build it. Everyone agreed that a space station would be good, but not on what it was good for.

President Kennedy's decision in May 1961 to send Americans to the Moon put the station on a back burner, since there was insufficient money for both projects. But even as work on Apollo moved toward its crescendo, NASA's space station designers quietly decided on a modular concept—a giant Tinkertoy—that would be built and supplied by a winged vehicle shuttling back and forth from Earth to orbit. Unlike the Mercury, Gemini, and Apollo astronauts, who parachuted out of space as though they were leaving the scene of an accident, those on the shuttle would return home in a more dignified gliding airplane.

When President Nixon approved the shuttle in January 1972, the station was postponed again. The heart of the problem had to do with a debate that continues to this day: Are people necessary for space exploration, or can machines work better and cheaper by themselves? Critical also was a budgetary pie of finite size. These key factors created a formidable number of enemies for the station.

There was no doubt where most of the science community stood. Scientists opposed the station because they felt its huge expense would drain money from much more useful research programs. In 1983, even as Reagan pondered whether to approve the station, the National Research Council's prestigious Space Science Board urged NASA to postpone it for at least 20 years because there was "no scientific need" for a station in space. But the scientists lost to NASA's persuasive administrator, James M. Beggs, who convinced Reagan that the station was "the next logical step" after the shuttle and that, furthermore, it would embody America's leadership in the world and stimulate free enterprise. Nevertheless, opponents in Congress and elsewhere continued to hammer at the station throughout the 1980s and well into the '90s. In 1991, the 40,000-member American Physical Society turned thumbs down on the station, saying that a vigorous space science program could be accomplished without it. The following year, representatives of a number of foundations and societies told Congress that there was no evidence that medical research in space helped patients on Earth. Dr. David Rosenthal of the Harvard Medical School attacked the notion that either cancer or AIDS could be combated any better in space than it could on Earth.

The station had its closest shave in 1993, when it survived a challenge in the House of Representatives by only one vote. But intense lobbying and a publicity blitz by industry, plus arm-twisting by Vice President Al Gore, finally turned the tide in 1994, when federal lawmakers voted 278 to 155 to provide $2.1 billion in funding for the following fiscal year.

By 1994, many scientists had changed their minds about the project and either weighed in for it or simply sat out the debates that raged in Congress. The change came for several reasons. That year, NASA shifted its public relations focus away from science, concentrating instead on jobs and cooperation with Russia. The science plan hadn't changed, but the project now offered a lower-profile target for opponents in the science community, who didn't want to take on the jobs issue or seem to impede closer relations with Russia.

Many scientists were also afraid to challenge the president, members of Congress, and the aerospace industry. As Columbia University astrophysicist David Helfand explains, defying NASA on its pet manned programs can bring reprisal to space scientists where their own grants are concerned.

Today, with the first launch approaching, at least one sector of the scientific community is excited about the prospects for research aboard the station: the 900 scientists with an interest in microgravity who are preparing experiments to send to the station. There, these experiments may shed light on the behavior of plants, animals, protein crystals, flames, molten metal, and other substances in the absence of gravity. There's no guarantee that these experiments will lead to anything as dramatic as a cure for cancer, although NASA and Boeing have raised that possibility. Yet in science, sometimes the best discoveries are the unexpected ones. "Serendipity will rear its lovely head," says Kramer, the former Lockheed scientist. "We'll find 10,000 things to do on the station that nobody's thought of or even imagined."

And yet even today, many scientists outside the field of microgravity research continue to criticize the station. The research planned for the station is "good science, but it's not science that would ever justify an expenditure like the space station to do," says Robert L. Park, a University of Maryland physics professor and the director of the Washington, D.C., office of the American Physical Society. Park is among those who believe that science would be better served by less costly robotic missions. "Robots improve every year, while humans have basically been the same for 35,000 years," he says.

For its part, NASA hopes to bring down the cost of the station by attracting paying customers. Some 30 percent of the research space in NASA's laboratory facilities has been set aside for commercial users such as pharmaceutical and biotechnology companies. Rep. Dana Rohrabacher of California, who chairs the House space subcommittee, has even suggested that the station be used for advertising and Hollywood filmmaking—but commercialization is being stalled by bureaucratic red tape. Bill Wisecarver, the exasperated director of commercialization for the station, complained last December that NASA's own space experts, grounded as they are in science and engineering, just "don't understand how the economics of the private sector work."

NASA is also contending with Russia's painfully slow work on the station's Service Module—the all-important power, control, and habitation structure from which the rest of the structure expands. The Service Module is already more than 60 days behind schedule. The cash-starved Russian Space Agency is also supposed to provide the propellant module and the resupply operation.

The Russian agency and its Energia and Khrunichev contractors desperately want to be viable partners in this international enterprise, not only to ensure their own survival and help rebuild their nation's crippled economy, but because space technology remains Russia's crown jewel and deepest source of pride. At the same time, the Russian legislature's unwillingness to honor the country's financial commitments to the station reflects a preoccupation with problems closer to Earth. "The biggest political risk," Boeing's Stone frets, "is dealing with the Russians."

"If the Russians pull out of the Service Module, yes, we can build the space station," says Gretchen W. McClain, NASA's

"Serendipity will rear its lovely head," says Kramer, the former Lockheed scientist. "We'll find 10,000 things to do on the station that nobody's thought of or even imagined."

director of space station requirements. But if they renege on the propellant and resupply operation too, she says, the station would be in serious trouble. "If they pull out altogether tomorrow, it would be a bad day on the Hill. I probably could see the program getting canceled."

But McClain remains optimistic. "It's been a battle and a struggle, but what's intriguing to me is that it is a concept that was started years ago and it has continued to survive a lot of challenges and upscaling and downscaling. For the first time," she says, "we've moved off of a paper study, which this program was on for years, and you're seeing hundreds and hundreds of thousands of pounds of hardware being completed."

Meanwhile, public support for the station appears solid. In a random telephone survey of 1,510 Americans conducted by Yankelovich in late 1997, 28 percent "strongly favored" carrying out a program to build a permanently manned space station, while another 42 percent "somewhat favored" it. For many Americans, space exploration remains a grand adventure that is well worth the money.

"If the shuttle and space station program went away tomorrow, you'd see the unmanned program go away," says McClain. "I believe the heart and soul of space is the human element. People can relate to that. You've got the young kids thinking, 'God, I'd love to be an astronaut. Wouldn't it be great to be up there, to touch and feel,' and I think that's part of the excitement about it."

For many Americans, space exploration remains a grand adventure that is well worth the money.

Cosmic Clunker[4]

It was not only pleasingly nostalgic but genuinely inspirational to behold John Glenn arcing into orbit again last month, reminding us that neither individual courage nor Kennedy-era dreams need ever end. But what comes next for the space program is unlikely to inspire. Due late in November is the first hardware launch for the space station, an expected seven-year construction effort. With flimsy scientific justification and an exorbitant price, the space station is a cosmic clunker.

Estimated annual bottled water cost for the station: $817 million. Is this a defensible use of public funds?

NASA's space station is a true international effort, as evidenced not only by European and Japanese roles in the project but also by the stunning fact that the facility's first component will be Russian-built and will thunder skyward aboard a Russian rocket. A decade ago, who could have imagined that? But the bulk of the project is NASA's: most management, manufacturers, launches, and crew will be American, along with almost all the funding. A few weeks ago, Bill Clinton approved an emergency subsidy of $660 million for Russia's space agency to keep it from going out of business, while NASA hardware sits in its clean rooms being prepped. This money is strictly a transfer payment from U.S. taxpayers to Russian aerospace contractors. Who could have imagined that, either?

The obvious concern about the station project is its immoderate cost. Administrator Daniel Goldin has done a wondrous job of reviving the space-probe side of NASA (think of that little, lovable Mars lander) but has dead-ended against the "manned" space lobby, which channels huge sums to facilities in favored districts in Alabama, California, Florida, Maryland, Ohio, and Texas. Today the station's sticker price is $21.3 billion, but $40 billion is more realistic when launch costs are taken into account. Then, once the Tang begins to flow and the station becomes operational, there'll be servicing costs. Space-station crew members, for example, will require several gallons per day of fluids for drinking and hygiene. At space shuttle payload prices, four gallons of water cost about $320,000 to place in orbit. (This is a conservative calculation.) So there will be the space-station crew, floating in an orbital hotel, consuming $320,000 worth of bottled water per person each day. Estimated annual bottled water cost for the station: $817 million. Is this a defensible use of public funds?

---

4. Article by Gregg Easterbrook. From the *New Republic* p6 + , Nov. 30, 1998. Copyright © the *New Republic*. Reprinted with permission.

It might be if the space station served some pressing scientific need, but it doesn't. The U.S. science community is close to unanimous in opposing the space station, which will sponge appropriations away from other, higher priorities. Space research itself is far more cost-effective when done by satellites and probes, several of which can be launched for the price of a single space-station resupply mission. The only research the space station will excel in is long-term observation of the physiological effects of space. That's fine, but is data on this abstract question really worth $40 billion when earthbound requirements for medical research are taken into account?

Collaboration with Russia might justify the project. Visiting mission control in Houston a few years ago, I was amazed to count the number of Russians wandering the halls of what was once a cold war alcazar. Goldin, a former executive of the Strategic Defense Initiative, told me, "During the cold war, I worked on weapons intended to destroy the Soviets; now, it is my moral obligation to show the world that the United States and Russia can cooperate peacefully on projects that benefit everyone." No one could fail to admire that sentiment.

In an unnoticed technical decision, Goldin stipulated that the new station be hung in an orbital "inclination" that is convenient to the Russian cosmodrome but all wrong for Cape Canaveral. The upshot of this decision is that the small number of Russian flights to the space station will cost as little as possible, while the payloads of NASA launches will fall, adding billions to the cost on the U.S. side: a huge, disguised gift to the Russians. Goldin's gesture here is a generous one, but it hinges on the cold war assumption that, when it comes to relations with Moscow, money is no object. NASA and Russia might, instead, merge their planetary programs, achieving the desired positive symbolism at a fraction of the cost of a space station.

The strongest objection to the station is that it places the cart before the horse. Someday, space stations might be desirable, along with even grander projects in space. Today, with each space shuttle launch costing at least $400 million, it's fiscal lunacy to be building a 475-ton chalet above the atmosphere. Instead, NASA's focus should be on developing new systems that cut the cost of access to space.

NASA's striking weakness is that the rockets it launches are almost as geriatric as John Glenn. The Atlas, Delta, and Titan boosters employed for satellite launches were all designed in the 1950s, before microelectronics, composite materials, or even pocket calculators. And the space shuttle is rather older

The Atlas, Delta, and Titan boosters employed for satellite launches were all designed in the 1950s, before microelectronics, composite materials, or even pocket calculators.

As it stands, there is more than a little chance that the station will be half-built and then canceled, a clump of powered-down modules circling overhead as a cislunar white elephant.

than it seems, its basic design commitments having been made back in 1972. Gleaming as a technical success, the shuttle has been a budgetary fiasco: Most launches cost considerably *more* per pound of payload than launches aboard throwaway rockets. No rational planner would use the shuttle for anything other than space-station crew rotation flights—a restive point for NASA, since, by all appearances, the space station is being built primarily to give the shuttle something to do.

NASA has fresh launcher ideas in mock-up, among them the X33, a "single-stage to orbit" machine that would rise to space in one piece, just like the rockets in "Buck Rogers" serials. Other ideas for advanced, lower-cost launchers are bouncing around the space agency and industry. If the X33 works, it might make access to space relatively affordable and routine, similar to a very fancy airplane flight. Then grand goals such as space stations or Mars travel could make sense.

The intelligent course is to postpone the space station indefinitely, diverting funds to basic rocketry research until the X33 or some other new affordable launcher is coursing the ionosphere. As it stands, there is more than a little chance that the station will be half-built and then canceled, a clump of powered-down modules circling overhead as a cislunar white elephant. NASA might be about to launch the first components of its own political demise. Good thing for Glenn he'll be off the flight list by then.

Building a Space Station Still Makes Sense[5]

When NASA was seeking White House approval to begin developing a space station in the early 1980s, it billed the station concept as "the next logical step" in space development. Finally, almost two decades later, the hardware phase of that "logical step" has begun. It is worth recalling, in the face of disputes about cost overruns, schedule slips, the Russian partnership and scientific payoffs, why building a space station made sense in the first place.

There have been many reasons why the program has survived the twists and turns in its evolution. Logic has not been foremost among them; the program from its inception until today has been supported primarily on international and domestic political grounds. Underpinning the politics behind the station are two logical reasons that lead toward development as the correct next step in human space flight.

The first is that if we intend to continue to send humans into space, they need somewhere to go. For three years, 1969–72, that destination was the Moon, but since then there certainly has been no will to continue to go back there or anywhere else away from Earth orbit. Space shuttle missions have been in essence brief camping trips to a nearby wilderness. After all, it is worth remembering that the shuttle was originally proposed only as the transportation vehicle to get people and supplies to and from a permanent orbital station, not as a surrogate for such an outpost: the shuttle can finally assume that role as station assembly proceeds.

The Soviet (now Russian) space program for the past quarter century has had an orbital destination in the Salyut and Mir stations. In that sense, that program bore more logic than we displayed in the U.S. by trying to make the shuttle an orbital laboratory—which it was not designed to be.

The space station represents a logical step if there is an intent to continue sending people into space. But why orbit humans in the first place? Answering that key question—whether long-duration human presence in Earth orbit has substantial instrumental value—provides the second logical reason for building an orbital lab.

Scientific critics of the station, adapting what seems to be a rather unscientific approach, have decided—in advance of its being performed—that multidisciplinary research in a

It is worth remembering that the shuttle was originally proposed only as the transportation vehicle to get people and supplies to and from a permanent orbital station, not as a surrogate for such an outpost.

---

5. Article by John M. Logsdon. From *Aviation Week and Space Technology* p78 Nov. 30, 1998. Copyright © 1998 *Aviation Week and Space Technology*. Reprinted wtih permission.

well-equipped laboratory operating in a microgravity environment will have no substantial payoffs. To me, it makes more sense to wait for those research results before deciding they do not justify the investment to obtain them. Perhaps there will be no substantial discoveries or other results justifying the costs of supporting humans in orbit; the station would then become a dead-end project in scientific terms, although it might still serve as a vehicle for U.S. leadership.

It is possible, however, that there will be significant payoffs from station research, leading to substantial growth in public and particularly private sector interest in orbital activities involving human presence. Perhaps in 20 years we will look back and wonder why there was any doubt that having an initial space station was worthwhile. This, of course, is a very optimistic scenario, but to me it is no less plausible than an *a priori* judgment that space station research will have no significant value. We should build and operate the facility, and let the results speak for themselves.

I have carefully avoided using the designation International Space Station (ISS). There certainly might have been better paths to creating a space station than the one that has been followed. No one would recommend the program's history as a case study in how to design and manage a major technological undertaking, and there is much to criticize in the current station plans. But looking backward is not productive.

In a Panglossian world, the ISS is now the best of all possible space stations, since it is the only one being built. There will be continuing problems during its construction, and the program's lack of robustness and dependence on the shuttle and in-space construction are particularly worrisome. At this point, however. there is no real choice but to move ahead and hope that nothing catastrophic happens. A combination of managerial excellence and good luck is to be hoped for to get the station up and running.

Meanwhile, watching the ISS come together will be exciting. People often watch large construction projects on Earth with fascination; it should be even more fascinating to watch the orbital equivalent.

It is possible . . . that there will be significant payoffs from station research, leading to substantial growth in public and particularly private sector interest in orbital activities involving human presence.

IV. Private Enterprise and Space Exploration

Editor's Introduction

In Stanley Kubrick's *2001: A Space Odyssey*, a Pan-Am shuttle plane is seen coasting towards a space station. Once inside, one of the film's characters is seen going into a Bell phone booth. While such overt commercialism has not yet arisen in space travel, it is fast approaching on the horizon. The involvement of private enterprise in space exploration truly began in 1976, when NASA launched the first direct-broadcast television satellite. Initially, private industry was hesitant about testing the waters in such a new endeavor. Once NASA proved such launches could be profitable, corporations moved in. In this last decade of the 20th century, the government has completely left the business of launching satellites. According to Peter Landesman, writing for the *New Yorker*, over 1,700 communication satellites will be orbiting the Earth by 2006, the majority of them sent up by the private sector.

In keeping with its new mantra of "faster, better, cheaper" missions, NASA has begun allowing academic and industrial investors to join unmanned missions. The Near Earth Asteroid Rendezvous (NEAR) probe is the first of these missions. In essence, the space agency's partners run the mission and NASA buys the scientific information gathered from it. NEAR's mission was to study Mathilde and Eros, two asteroids within our solar system's asteroid belt, and relay information back to Earth about them. The probe made contact with Mathilde on June 27, 1997, and with Eros on December 23, 1998. This mission marked the first fledgling steps made by the private sector into the investigation of extraterrestrial bodies.

Currently, the first privately funded mission to an asteroid is being prepared for an April 2001 launch. The *Near Earth Asteroid Prospector* (NEAP) is revolutionary on many levels. The probe is being launched not by NASA, but by a private company, the Space Development Corporation (SpaceDev). If successful, it will be the first unmanned probe to touch down on an asteroid, specifically the asteroid Nereus, which the company's owner, James Benson, will claim as his private property. Further, it is the first mission in which scientists are paying a private company to conduct experiments, instead of the reverse. Beyond SpaceDev, other private space exploration companies are forming at a rate which suggests that if NEAR is a success, a revolution regarding how humanity journeys into space could be right around the corner. Already there is talk in many circles of mining the Moon for its mineral potential, or developing it as the ultimate get away.

The articles in this section describe a number of projects on the horizon that involve members of the private sector. The first of these articles, Peter Landesman's "Starship Private Enterprise," published in the *New Yorker*, discusses James Benson and NEAP in great detail and evaluates the successes of the *Lunar Prospector*, a probe launched in January 1998 by a private company and designed to orbit the Moon and relay data. Already the probe has confirmed the existence of water on the Moon—something that scientists at NASA had speculated about for years but were unable to investigate. The author also remarks on the state of current space law regarding in individual's right to own celestial bodies as well as a number of other topics relevant to private enterprise in space. The second article, "Buck Rogers, CEO," by David Schneider for *Scientific American*, details the *NEAP* mission as well as a number of other projects, including Luna Corporation's ambitious project of attempting to get to the Moon by 2001. In the third article, "Looking for Money on the Lunar Surface," published in *Newsweek*, Theodore Gideonse describes the various companies wishing to invest in the Moon, in everything from strip-mining to setting up hotels to developing new forms of cheap solar energy that could be beamed directly back to Earth. Finally in "Reach for the Moon" for the *New Statesman*, David Whitehouse argues for a return to the Moon. In this article, Whitehouse gives a brief history of the Moon and notes the benefits of developing a Moon base. Noting that oxygen makes up half the weight of Moon rocks, he goes on to remark that besides the obvious reasons for needing oxygen on a Moon base, oxygen can be used as a rocket fuel. He argues that returning to the Moon "is the only real way to get the public interested in space again. The space shuttle doesn't go anywhere, except round in circles, the space station is worthy but dull and a trip to Mars is just too expensive. To capture the public's imagination it has to be a trip to the Moon." Many entrepreneurs are banking on it.

Starship Private Enterprise[1]

Early next year, one of the most extraordinary capitalist dramas of the late 20th century will begin in earnest. In a warehouse outside San Diego, engineers and skilled workers will start putting together a new kind of spacecraft, the unmanned *Near Earth Asteroid Prospector* or NEAP. Then, in April, 2001, the craft will be launched—not by NASA or some other government entity but by SpaceDev, the world's first commercial-space-exploration company. Thirteen months later, after flying two and a half million miles, the spacecraft will rendezvous with Nereus, an asteroid whose irregular orbit around the sun will bring it into the Earth's cosmic neighborhood early in 2002. If all goes well, the spacecraft will conduct experiments for scientists who have paid to put instruments on board; it will also land instruments on the asteroid itself, something the like of which has never been attempted before. About a tenth of the way to its meeting with Nereus, *NEAP* will fly past the Moon and drop a probe to scrutinize the lunar surface with the same sort of camera *Pathfinder* used to explore the surface of Mars in 1997. Above all, though, the expedition will mark the beginning of an extraterrestrial industrial revolution fueled by a combination of scientific ingenuity and commercial drive that may look, at first glance, like space piracy.

Jim Benson, the founder of SpaceDev, is a 52-year-old software entrepreneur who has the ruddy looks of a Midwestern farm boy and a closet of identical blue suits. (He grew up in Kansas City; his father was a cattle rancher, a stockbroker, and an adman.) Since he established the company, he has recruited many of the world's leading authorities on space to work on his design-and-management team, either as consultants or as full-time employees. Among them are a one-time mission planner for the Apollo manned space program; the lead investigator on the Lunar Prospector mission, which last January discovered signs of water on the Moon; and a former top NASA official. NASA administrator Daniel Goldin is also an enthusiastic supporter of the new venture. He has distributed some of the funding NASA earmarks for private enterprise so that universities and scientists can meet Benson's price. (Running an experiment on *NEAP* will cost 10 to 15 million dollars.)

---

1. Article by Peter Landesman. Originally published in the *New Yorker* p 178-185 Oct. 26 & Nov. 2, 1998. Copyright © 1998 Peter Landesman. Reprinted with permission.

Benson's project raises serious questions about the ownership, exploitation, and regulation of outer space in the next century.

Benson's project raises serious questions about the ownership, exploitation, and regulation of outer space in the next century. The project could also make members of his team extremely rich. Some asteroids are flying mountains of stainless steel and precious metals such as gold and platinum. Other big space rocks, called carbonaceous asteroids, offer conditions—carbon, water, moderate temperatures—that could have once supported, or might still support, primitive forms of life. Once *NEAP* makes contact with Nereus, Benson plans to declare the asteroid his private property and assert his right to mine it both for minerals and for scientific data. "No longer is outer space the exclusive realm of engineers and scientists," he told me recently. "Multidisciplinary people are involved now—doers, entrepreneurs, real-estate developers, investment banks." Paraphrasing an old Steven Jobs Apple Computer slogan, Benson summarized SpaceDev's mission as "Space for the rest of us."

SpaceDev's East Coast office, where I talked with him, is a sparsely furnished suite of rooms situated in a tower of reflective glass amid other towers of reflective glass in Arlington, Virginia. The company's main headquarters is in San Diego, but Benson recently opened this office, in the suburbs of Washington, D.C., to be near the embassies of countries that may someday buy a ride on his space bus. He seemed perplexed that so few people grasp the importance of the project. "Outside the box the space community lives in," he said, "the world has no idea what is about to happen."

This is Benson's stump speech, and he delivers it almost daily, with the automatic exuberance of a politician, to potential investors, NASA officials, and space enthusiasts. He mixes the nerdy boosterism of high-tech capitalism with the urgent idealism common among aerospace visionaries. Space exploration is not just about spaceships and flags and footprints, he tells his listeners with his eyes alight. It's about finding alternatives to life on Earth.

"Buckminster Fuller said that Earth is a spaceship," Benson said. "We have our own life-support system. It's the oceans, it's all of life. And it's all interrelated. It's a spacecraft with a built-in life-support system, and, piece by piece, we are dismantling it as we fly through space. How long can we keep it up?" He added, "If we don't drown in our own [mess], we'll get whacked"—by an asteroid. The kind of space exploration Benson envisions, he explained, "may offer solutions in both respects, moving resource exploitation into space, where there isn't any life and never will be any— onto asteroids—and creating the industrial infrastructure to allow settlements for humanity to survive the ultimate catastrophe. . . . Everything else is just wishful thinking. When you see *Deep Impact* or *Armageddon*—I mean, planting nuclear weapons on those things? It'll never happen. We don't have a clue what to do about them. And until we get up there and start working with them, and making a living off them, we won't have a clue."

This summer's catastrophic explosions of two rockets shortly after liftoff—Boeing's *Delta III* and Lockheed's *Titan IV A*—were expensive reminders that space is a risky business. More than a billion dollars' worth of technology blew up, and the market for commercial satellite launchings was dealt a severe setback.

Still, the numbers that Benson puts on the table are persuasive. It will cost SpaceDev less than 50 million dollars to send *NEAP* to Nereus, and the trip could generate revenues of at least 120 million dollars in research and data fees, which would yield a profit of more than 70 million dollars. In addition to being lucrative, the payload of lunar and asteroidal experiments that Benson hopes to assemble will be historic. For the first time, one space mission will do research for university scientists, for NASA, for private corporations, and for Japanese and European space agencies. Benson also intends to design three experiments of his own and sell the data to NASA, whose policy mandate is to buy all commercially available space science.

Benson envisions the outer shell of *NEAP*—a craft of roughly nine cubic yards, which looks in computer renderings like a squared-off bucket awkwardly winged with a pair of solar flaps-covered with the logos of companies like Microsoft ("Where do you want to go today?") and Visa ("Anywhere You Want to Be"). An Israeli milk company has already paid Russia to shoot a commercial on Mir, its creaky space station. So it is not difficult to imagine *NEAP* hurtling into space tattooed like a NASCAR racer.

Benson is not an inventor, but an ingenious shopper who bought parts off the shelf. "It's not like I'm some crackpot scientist who has come up with a brand-new widget that is the greatest widget ever designed," he said. Between 80 and 90 percent of NEAP will be assembled from components that were developed and tested on previous NASA missions—including the flight computer, the solar panels, the rocket thrusters, and the spacecraft's skin.

If it all works as planned, SpaceDev will have revolutionized not only spacecraft construction but also the way in which space science is conducted. Cheaper, more efficient missions would mean that "you'll do planetary science and exploration in a better way," Lynn Harper, a co-leader for

Between 80 and 90 percent of NEAP will be assembled from components that were developed and tested on previous NASA missions— including the flight computer, the solar panels, the rocket thrusters, and the spacecraft's skin.

astrobiology at NASA's Ames Research Center, told me. "If you only get to go someplace once, you're going to hit the features of greatest interest with the fewest risks, because you're only going to have one chance to collect data. That means you're going to look for what you're pretty sure is already there. When you go multiple times, you can take higher risks, with the potential for higher payoffs. On comets and planetary bodies, that includes the search for life."

The closest thing to a test run of the Benson model was Alan Binder's *Lunar Prospector*, which was launched in January of this year and has spent five months orbiting the Moon, collecting data. The Prospector, a privately designed and managed project, was the most cost-effective lunar and planetary space mission ever flown. It performed its task flawlessly at a tenth of what many believed NASA would have spent. When Prospector started, most of my colleagues wouldn't touch it with a ten-foot pole," said Binder, a NASA veteran who is the principal investigator of the Prospector mission. "It was heresy—it was 'NASA will not like this.'" Binder's organization, the Lunar Research Institute, now has an informal, cooperative relationship with Benson. . . . Binder exclaimed gleefully, with a sweep of his arm. "I think NASA is the enemy and should be treated like the enemy. They threw away 25 years. We had Apollo. We not only had the capability of putting a man on the Moon, we had plans and the ability to have a lunar base by the 1980s. We threw all that away because of the Vietnam War, social issues. We went to the Moon politically. We wanted to show that we could beat the Russians. The public did not understand why we were doing this. But we dropped it dead. The exploration of space as a government program is just never going to do it. It's too costly. There's no payback. Science is wonderful, but the American public is not going to get excited about science. So until we learn how to do this commercially we're not going to explore the Moon the way we have to."

Binder pointed to the spending of 40 billion dollars for the projected International Space Station as the epitome of NASA ineptitude. It's a political football," he said, laughing. "The shuttle is no damn use. NASA justifies it by saying we need it for the future, to support the space station. But the space station is needed for what? Well, there's no answer. It's a catastrophe. It's how many years late, how many billion dollars over budget? When it's up there, it's not going to be extremely useful. The public is not interested in it. This is leading nowhere." Indeed, NASA will be requesting an additional 600 million dollars to, in effect, bail out the bankrupt Russian space agency, its main partner in the space station.

[The project manager of the Lunar Prospector Alan] Binder pointed to the spending of 40 billion dollars for the projected International Space Station as the epitome of NASA ineptitude.

Russia, in turn, has sold much of its allotted research time to NASA for an additional 60 million dollars. The space station, which began as a post-cold war project to unite the world aerospace community, has become a millstone for America.

NASA's top officials understand that they may no longer control the future of space exploration. Wes Huntress, NASA's outgoing associate administrator for space science, believes that Benson has the potential to be a historic figure—the Henry Ford of the 21st century. "He's the only one who has the vision," Huntress told me in his office last spring. "There have been many people coming through my door who say, 'I want to do a private commercial mission, here's what I want to do, now give me a hundred-million dollars. This is not a private commercial mission. Jim's first words were, 'I won't want your money.' That got my attention. All he wanted was the opportunity to compete."

Nevertheless, Benson's business plan relies heavily on NASA funds. His first customers are likely to be the scientists who receive grants from NASA's Discovery Program, which was set up expressly to help scientists pay private companies that can facilitate their experiments. Likewise, Benson will be able to sell his own space research only as long as the American, the Japanese, and the European space agencies continue to purchase data—and thus sustain a potential billion-dollar-a-year market. In a similar manner, Binder's Lunar Prospector mission, which was touted as the first successful commercial spaceflight, was in fact made possible by a NASA contract.

When I suggested to Benson that NASA will always have a technological advantage over private industry, because its research can be incubated away from the pressures of impatient stockholders and quarterly reports, he shrugged. "Starting in the next 10 years, as space commercialization takes off and the companies are profitable, and they get into competitive mode to do their own R&D to compete with each other, the need for NASA is going to start fading," he said. "We've been going nowhere for 20 years. The space shuttle is going in circles. Astronauts are playing with Slinkys."

Dan Goldin, the NASA administrator, doesn't necessarily disagree with Benson's critique. "If we set mediocre goals, we get mediocre results," he told me. "If we set goals not to have failure, we never learn. One of the problems we have here at NASA is that we're so afraid of failure we spend billions of dollars protecting ourselves against it." Goldin speaks of Benson's project with great enthusiasm. "This is a hell of a courageous thing to do," he said. "It will get people's attention focused on the frontier still to be conquered."

Wes Huntress, NASA's outgoing associate administrator for space science, believes that Benson has the potential to be a historic figure—the Henry Ford of the 21st century.

Goldin spoke with me in his NASA office. He wore cowboy boots, but his speech bent and stretched with the long vowels of his native Bronx. He is known for his gung-ho, shoot-first-and-ask-questions-later dedication to the space program. A table in his office was covered with plastic models of the agency's current projects—the X-33 rocket plane, pieces of the International Space Station, and various spacecraft, all white and gleaming and aerodynamic. Fiddling with a model, he said, "Look, all we do is make movies about Apollo. What we do in America is not write history—we watch history. And we're getting satisfaction out of entertainment based on historical fact. Opening the space frontier tells our kids that this is not about mediocrity—it's about taking risks. Maybe risking one's life. It's about failing and then coming back and succeeding. Space is more than entertainment. It is a projection of what we all want: some meaning in our lives."

Goldin is redrafting NASA's mandate to create a more streamlined agency that will focus solely on deep-space technology. By his own estimation, within 10 years NASA's gradual withdrawal from activity between Earth and the Moon will put 75 percent of all space efforts, commercial and scientific, into the hands of private citizens like Benson. Houston Mission Control was contracted out to a private consortium last month. The space shuttle is already managed, in part, by private contractors, and the International Space Station will soon be on the block.

NASA hopes that stepping aside and letting the more efficient private sector develop space can work as well as it did for communications satellites. NASA launched the first direct-broadcast TV satellite in 1976 to assist global cooperation in medicine. The private sector was unwilling to assume the initial risks of testing the new technology, but once the government had proved it could be profitable, corporations rushed in. NASA finally got out of the business in the early 1990s. By 2006, more than 1,700 commercial communications satellites will be in Earth's orbit. NASA's head of policy and planning, Alan Ladwig, had considered leaving for the private sector, but then decided that for now he is more valuable in government, where he can smooth the road for private corporations like SpaceDev. Still, he remains clear about his priorities. "A success by Benson will do more to jumpstart the economic return from space than sending a handful of astronauts who will plant flags on Mars," he told me. "The agency's role has shifted from being the implementer of the space program to being the provider of the technology that will allow the country to get into space. Our purpose is to do something new and different that industry will be able to use after we're done with it."

Goldin agrees. "I am going to focus on technology that is impossible. I am going to put up spacecraft three times the size of the Hubble, separate them by three or four football fields, and I'm going to have to know their exact position to within a billionth of a metre. Then I'm going to have to physically place them to within one inch, and hold that one-inch distance at half a billion miles. On top of that, I'm going to send out robots and train robots to change broken equipment, and communicate with the robots over that enormous distance. No corporation can do that. We will do things that no one can even conceive of doing."

Jim Benson's mission to commercialize space began a little more than two years ago. At the time, he was 50, financially secure, and had retired to a new home, in Steamboat Springs, Colorado. But his convictions about space as a source of great hope and great fear for mankind date back to the early seventies. As a college student, in the sixties, Benson had become infatuated with the earliest generation of computers, and had taught himself to program. In 1972, he took a computer job in Washington, D.C. Soon after he arrived, a neighbor gave him a copy of *The Limits to Growth*, a study for the Club of Rome, which warned, in apocalyptic terms, of global warming and resource depletion. Until then, Benson had considered himself a default conservative," essentially apolitical. He had been alienated by Vietnam and Watergate. *The Limits to Growth* gave him a vision of the interconnectedness among land, sky, water, and man. He quit computer programming, took the Civil Service exam, and went to work for the Solar Energy Division of the United States Energy Research and Development Administration, as the Department of Energy was then known.

At ERDA, Benson earned a reputation as a crusader. He managed an impact study of the Ford Administration's National Energy Plan, which concluded that the government was ignoring new technologies such as solar and wind power while subsidizing oil, gas, and nuclear power. Benson became an adviser to the Presidential campaign of Governor Jimmy Carter, whose environmental goals were more in line with his own, and he was one of a half-dozen environmentalists who helped frame the Carter Administration's energy policy.

But his science fell under suspicion at what by then had become the Department of Energy. A high official at the department during Benson's time there, who asked not to be identified, described Benson as a zealot who slanted data to favor alternative energy sources. Benson denies this. "People who were unhappy with those objective results were party-liners, who accused me of not being a team player, because the results did not agree with their preconceived and politically derived conceptions," he said. In any case, Benson was asked to resign. His boss in the Solar Energy Division, Hank Marvin, recalled that it was clear, even then, that Benson was too driven to survive in a bureaucracy. "He is just what I thought," Marvin said, "reaching one step beyond what we thought we could reach in our lifetime."

Benson became an independent computer programmer, and in 1984 won a a contract to write software for new federal procurement regulations. Among other things, his company, Compusearch, designed the algorithm that in microseconds can find particular words and phrases in huge bodies of text—one of the precursors of the Internet search engine. He sold Compusearch in 1995 for several million dollars, and in 1996, straining against the confines of inactivity, began to revive his childhood fascination with outer space. He immersed himself in aerospace literature, compiled a list of what he considered the 10 most brilliant minds in the industry, and E-mailed them one by one. What, he wanted to know, was there left to do? What would be interesting and useful? What could make money?

One man—Steve Ostro, of NASA's Jet Propulsion Laboratory, who is one of the world's leading radar astronomers—wrote back. Using radar, Ostro had recently studied the asteroid 1986 DA, and determined that it was the old metallic core of a minor planet that had once existed between Mars and Jupiter—essentially a mile-wide mountain of stainless steel and precious metals, mostly gold and platinum. Ostro advised Benson that a scientist at the United States Geological Survey was already putting together an economic model of how milling asteroids might affect the global metals market, and he directed Benson to John S. Lewis, the co-director of the Space Engineering Research Center at the University of Arizona, who had written a book called *Mining the Sky*. Lewis convinced Benson that in the right political and economic climate the industrialization of outer space was not only imaginable but practicable with existing technology; that a new, off-Earth economy was waiting to be ignited; and that an extraordinary amount of money was waiting to be made.

"We are in a Darwinian mode, Lewis told me. "Today, everything is an intelligence test. If you find yourselves simply perpetuating the status quo, then you are dying. . . . The technical problems are secondary. This country has a huge reservoir of expertise in launching space missions, huge reserves of people laid off by the space and defense industries, people chomping at the bit." Lewis went on, "The supposed limits of growth are based largely upon false assumptions. . . . Because the space age was developed between mortal enemies—capitalism and Communism—it got us into a deep rut. It's great to jump-start space by having it as a national goal. But when the government can't let go, that kills it. In the Second World War, the government dominated all kinds of aircraft. One of the reasons for America's powerful position in the modern world is that the aircraft industry successfully stepped out from under government domination. Space could have done the same thing, and it still can, but the government has bitterly opposed losing its monopoly."

Benson had heard his calling. He moved quickly to incorporate SpaceDev, installed himself as chairman and C.E.O., and acquired Integrated Space Systems, a San Diego aerospace engineering company that had been doing spacecraft design and analysis work for Lockheed. SpaceDev has since purchased Space Innovations Limited, a spacecraft-and-satellite-products company, and the technology patents of American Rocket. Today, SpaceDev is the only company in the world with the ability to design its own space mission, build its own craft, and get it up in the air with its own rockets.

Among the top scientists that Benson assembled was Tony Spear, a 36-year veteran of NASA, where he served as project manager for the Mars *Pathfinder*. Spear is an articulate and clear-eyed spokesman for space exploration, and he is a bridge between its past and its future. He says that aerospace equipment, like computer hardware, has become exponentially better, cheaper, and faster over the last 15 years. "Benson is on the verge of a major breakthrough in spacecraft technology," Spear told me. "In five years or so, spacecraft costs are going to plummet to the point where whole spacecraft and associated ground systems are going to cost less than 10 million dollars. Universities can have their own spacecraft. Mexico can have a spacecraft. But there's a fine line between being merely low cost and irresponsible versus extremely cost-effective." Spear added, "There is an established way for you to do deep space exploration, and that establishment has a lot of history and trial and error. . . . In the process of cutting out the fat, you can cut into the meat and make mistakes." When he talked about Benson, Spear, like many of the space veterans I spoke with, could barely contain his excitement. Scientists in general are hoping and praying that he makes it—that he opens up low-cost oppor-

Today, SpaceDev is the only company in the world with the ability to design its own space mission, build its own craft, and get it up in the air with its own rockets.

tunities for scientists to jump on board without all of the paperwork . . . that they have to go through right now," Spear said, and then paused, as if making an effort to rein in his enthusiasm. He was most impressed, he went on to say, by Benson's insistence on identifying the most successful space missions and recruiting the men and women who had made them so. But the principal attraction for Spear is not science or recognition. "If I'm going to work for him," Spear told me bluntly, "he's going to have to make me very rich." At present, though, SpaceDev stock, which is traded on the NASDAQ, fluctuates wildly, reflecting the uncertainties of both a volatile market and a brand-new industry.

Money matters to Benson, too, but so does the vision. For all his talk of the bottom line, what he sees first when he looks at space is a sanctuary from mankind's ecological despoliation of the home planet. He loves to talk about how globalization has made Earth psychologically small, and how, as its population approaches 12 billion, there will be nowhere to go but up. He believes that only the drive and ingenuity of the private sector will get us there. Yet he also knows that there are risks in commercializing and unresolved questions of ownership and regulation. Existing space law was left purposely vague and toothless by its major signatories: the 1967 United Nations Outer Space Treaty does not address the issue of natural resources in space.

This is the one topic that makes Benson testy "Look, I've got three pieces that I'm juggling," he told me. "The business part of this has to be absolutely solid, impeccable. There's got to be a return on the investment or it's never going to happen. That's absolutely Rule Number One. . . . Then, there's the wonder of space and the enormity of it, and how little we know and how much more there is to learn." The testiness gave way to enthusiasm as he continued, "The third part of it is that I will pursue them"—resources in space—"but I won't be cheapened by gold mines in the sky." But, I asked, what is to prevent these extraterrestrial colonies from resembling Dickensian London, or the squalor of gold-rush California? "These colonies are going to grow like boom towns," Benson admitted. "There is going to be no planning. It will be an economic workhouse. You're going to wind up with prostitutes in space and blue-collar workers and office workers, and people are going to die, they are going to be killed, and we are going to find places to squeeze people into some tuna can up there."

When I asked Dan Goldin if he had any concerns about a private corporation launching a business in virgin space, he replied, defensively, with a question of his own: "Is it greed when someone wants to create opportunity? There's a fuzzy area as to who owns what. I don't see it as greed—I see it as man going for it, a real possibility for our progeny." Reddening, and clasping and unclasping his hands, he continued, "There are people throughout history who know with certainty what can't be done. They know it, feel it in their hearts. These are not the people who created the opportunity. These are not the pioneers—the ones who made fortunes, lost them, and made them again. These are not the people who died trying to make something happen. Jim Benson is necessary for progress. . . . He has the unmitigated gall to step up and say, 'I'm going to do it. Stop me if you can.'"

After a speech Benson gave last spring at a conference of aerospace engineers in Albuquerque, I was taken aside by a man named David Livingston. Livingston, a business consultant from San Francisco, makes his living in import-export, but has also been one of the first businessmen to think seriously about corporate behavior in space. Livingston, who is 43, diminutive and friendly, had been listening to Benson with a combination of suspicion and grudging admiration. "What comes off this Earth is us, a reflection of us," he said. "Someday, colonies will be self-supporting. Someday, they may be their own states or their own nations, and may be connected to Earth only in the way that Puerto Rico or Guam is connected to the United States. They are going to be extensions of us, with sovereignty. Is our greed in business so ingrained in us now, is our corruption, our lack of values, our lack of spiritual values, our lack of humanity? Is that the side of us that is going to go out into space? Is it profits, and damn everything else?"

In June, at SpaceDev's headquarters, in San Diego, I sat in on a brainstorming session to which Benson had summoned eight current and prospective advisers in order to weigh, among other things, the advantages of flying a safer (i.e., longer and less ambitious) mission versus a riskier, quicker one. Don Yeomans, who is the senior research scientist at NASA's Jet Propulsion Laboratory, said he wondered whether SpaceDev should play it conservatively the first time out—an asteroid flyby rather than an actual rendezvous. Everyone listened carefully. Yeomans is a crucial part of the venture, an ambassador to the aerospace cultures in government and academia. "If we go flying by an asteroid, the public's going to be yawning," Benson said, tapping his pencil in irritation. "If we're successful in this, we can do anything." But Yeomans continued to voice objections to everything from the flight path to Benson's choice of asteroid. Jim Arnold, a former Apollo mission planner and the director of the California Space Institute, and one of Benson's early champions, shook his head and muttered, "Look, nobody's ever done this before. You haven't done this before. We haven't done this before. Everything that gets done gets done for the first time."

"In or out?" someone asked Yeomans. Yeomans didn't hesitate. "Oh, yes. I'm in," he said, nodding. Phil Smith, SpaceDev's chief operating officer, leaned over to Bob Farquhar, the mission designer of the Applied Physics Lab at Johns Hopkins and the mission manager of NASA's CONTOUR mission, which is due to launch in 2002. In 1985, he designed the first mission to a comet. "How many science decisions have you made based on the stock market?" Smith asked, smiling ruefully. Within an hour after the meeting, Farquhar was talking about signing a contract with Benson, though in the end they failed to come to terms.

"Nobody thought you could break the four-minute mile until somebody did it, and then it became fairly commonplace," Benson said to me later. "It was more psychological than anything. If we break this four-minute mile of space, go out and make some money doing something practical that is needed in an existing market, all hell is going to break loose. . . . I don't know what's going to happen, but at least I will have the satisfaction of knowing that I started something."

Buck Rogers, CEO[2]

Not so long ago the idea of mining the Moon or asteroids belonged strictly to science fiction. Then more visionary thinkers began seriously considering how tapping the wealth of materials in space might open the solar system to commerce sometime in the 21st century. Now a few real-life entrepreneurs are planning ventures to exploit Earth's near neighbors over just the next few years. Curiously, the makeup of some current scientific missions suggests that a shift toward commercialization of space exploration is already quietly under way.

The Near-Earth Asteroid Rendezvous (NEAR) probe recently raced past the asteroid Mathilde, on its way to the asteroid Eros. NEAR is one of the so-called Discovery missions sponsored by the National Aeronautics and Space Administration. Discovery programs must cost less than $150 million (in 1992 dollars) and require no more than three years to develop. But there is a more subtle aspect. Rather than being a NASA mission per se, NEAR is being run by a set of academic and industry partners; NASA essentially just buys the scientific data.

This change in the way NASA is doing business may be creating opportunities for more obviously commercial efforts. For example, James W. Benson, a retired software entrepreneur, founded Space Development Corporation [SpaceDev] this past January with the intention of launching a private probe to another near-Earth asteroid. His interest in space was sparked in 1991 by an article in the *Washington Post*, which referred to a metallic asteroid as "an astronomical El Dorado." Initially, Benson planned to have his private craft fly to an asteroid and stake a mining claim, but he soon realized that the space probe could also carry scientific instruments and that he could sell the data. "We're going to be the first private exploration company," Benson asserts.

Benson has assembled a cadre of volunteer aerospace engineers to design his craft, *Near-Earth Asteroid Prospector*. In many ways, this group effort resembles the early stages of Lunar Prospector, a Discovery mission that will be sent to the Moon at the end of October.

Alan B. Binder, the principal investigator, explains that Lunar Prospector was originally "meant to be a demonstration of commercial viability." Binder and his colleagues tried

Rather than being a NASA mission per se, NEAR is being run by a set of academic and industry partners; NASA essentially just buys the scientific data.

---

2. Article by David Schneider. From *Scientific American* p 34-35 Sep. 1997. Copyright © 1997 *Scientific American*. Reprinted with permission.

David P. Gump, president of LunaCorp, is still hopeful about getting to the Moon before 2001 by looking outside the U.S. for support: "For the rest of the world, going to the Moon is a new thing."

to raise funds from private sources in the early 1990s. At the time, the project was estimated to require only $10 million in addition to launch costs, which were to be picked up by the Soviets using their powerful Proton rocket. But when a $4 million contract with Pepsico (for advertising rights) failed, the venture foundered. Yet *Lunar Prospector* was positioned perfectly to win the competition when NASA announced the Discovery program of economy space exploration. "I had a cheap mission that was ready to go," Binder recalls.

Binder, who is now retiring from Lockheed Martin, intends to mount privately sponsored lunar missions in the future and to sell the data obtained to NASA. Other lunar businesses hope to profit without depending on the space agency as their primary client. For example, LunaCorp in Arlington, Va., has teamed up with the Robotics Institute of Carnegie Mellon University in an effort to place at least two small roving vehicles on the Moon, which would then serve as the central attraction for a theme park. A dress rehearsal of the concept took place this summer: the Robotics Institute sent a small "rover" to roam Chile's Atacama Desert. Paying visitors to the Carnegie Science Center in Pittsburgh had the opportunity to drive the robotic vehicle remotely.

It is remarkable to think such ticket sales might be sufficient to fund a sophisticated mission to the Moon, but some in the business community are close to being convinced. According to William L. Whittaker, director of the Field Robotics Center at the Robotics Institute, his joint effort with LunaCorp almost achieved financial liftoff last year. They were negotiating with Walt Disney Company, among others, and nearly persuaded their would-be backers of the reliability of their chosen launch vehicle, the Russian Proton rocket, when the Proton carrying the Mars '96 probe sent its payload tumbling into the Pacific Ocean. That failure, Whittaker believes, gave the potential corporate backers cold feet.

David P. Gump, president of LunaCorp, is still hopeful about getting to the Moon before 2001 by looking outside the U.S. for support: "For the rest of the world, going to the Moon is a new thing." His company is now negotiating with a Japanese television network.

Perhaps it should not be surprising that people are seeking to make money through various entertainment schemes. After all, the current Discovery missions cost only about as much as some major motion pictures. But one business hopeful is also examining the possibility of "mining" the Moon for profit. Brad R. Blair, a geologist and mining engineer, created a company called Harvest Moon to establish just how profitable Moon rocks might be if sold more widely.

The idea came to Blair after discussions with David S. McKay, a NASA scientist whose former work on space resources has been eclipsed by his recent notoriety in claiming to have found evidence for ancient life on Mars. Blair was talking to McKay shortly after Sotheby's auctioned one carat of Moon rock for $442,500 in 1993 and realized that, extrapolating crudely, a kilogram of Moon rock would be worth $2.2 billion—far in excess of the cost of retrieval. The price of Moon rocks would surely drop if the supply grew, but overall revenue could still be enormous.

A properly authenticated Moon rock could become like "a rare mineral or a rare gemstone," according to Blair, who notes that a piece of lunar rubble brought to Earth several months ago in the form of a meteorite was offered for $200,000 per gram. At that rate, maybe a spaceborne El Dorado is really not so far away.

Looking for Money On the Lunar Surface[3]

Once *Apollo 11* landed in the Sea of Tranquillity and Neil Armstrong picked up a few rocks to prove he'd been there, the urgency that got NASA to the Moon vanished. Within a few years, the missions stopped. "The country basically made a decision that it was worth beating the Russians to the Moon but it was not worth it to do scientific exploration of the Moon," says Kent Joosten of NASA's Exploration Office. For years, humans went to the Moon only in science fiction. But now some people whose pockets are as deep as their eyes are wide are looking to the Moon again. This time, though, it's about profits, not patriotism.

NASA still thinks space is the final frontier, but like the rest of the shrinking government, the agency wants the private sector to pick up more responsibility—and more of the tab. NASA's plan is to put more emphasis on research and development. "NASA shouldn't be building the towns of the future, they should be building the highways," says Bettie Greber, the executive director of the Space Studies Institute in Princeton, N.J. In the future, NASA will probably act as the advance team for the entrepreneurs and big-name aerospace companies that have business plans in space as grand as, but more profitable than, Kennedy's mission of the 1960s. NASA will lead the scouting parties. Let others build the colonies.

The Lunar Resource Co.'s Artemis Project is one of the most ambitious plans. But it is so reminiscent of sci-fi fantasy that even LRC vice president Ian Randal Strock says that one of the main obstacles to finding investors is "getting over the giggle factor." LRC wants to send tourists to the Moon, set them up in luxury hotels and let them walk on soil "no one has ever walked on before." The Moon would be the most exotic vacation spot possible—and only a three-day, zero-G space shot away. Right now LRC is starting small— Moon mugs, Moon T-shirts and Moon calendars. It estimates the first tourist junket won't arrive on the lunar surface for 15 years or so, and the first retirement community will open 20 years after that "because that's when the founder of the project wants to retire," Strock says.

The Moon wouldn't just be a tourist destination. Some see it as the next Saudi Arabia. LRC and other firms have plans

Now some people whose pockets are as deep as their eyes are wide are looking to the Moon again. This time, though, it's about profits, not patriotism.

---

3. Article by Theodore Gideonse. From *Newsweek* 131 10 Mar. 9 '98. Copyright © 1998 Newsweek, Inc. All rights reserved. Reprinted by permission.

to mine the Moon's surface for a rare-on-Earth, abundant-on-the-Moon isotope of helium used for cheap fusion. And the potential for solar energy on the Moon is unlimited because it has no atmosphere to break through, and the days are long and bright. The cheap, unlimited energy could be beamed back to Earth as microwaves.

Sending up the tools to harness the sunlight and dig up the helium would be prohibitively expensive, so companies like the Shimizu Corp. want to strip-mine the lunar surface for, among other things, the raw materials to build the machinery on the Moon itself. Silicon exists in Moon soil in extremely high concentrations; it could be used for mirrors for solar-power generators, and there's plenty of iron for the wiring.

These raw materials could also be used to turn the Moon into the shipbuilding capital of the solar system. Building shuttles and satellites on the Moon would allow them to be launched from the low-gravity lunar surface. Less fuel, and thus less money, would be needed.

Lunar land will eventually be quite valuable. Buy now. For 18 years the Lunar Embassy of Rio Vista, Calif., has been offering 1,800-acre plots of land on the Moon and the planets for the low, low price of $15.99 (plus lunar tax). Whether it is legal to sell the Sea of Tranquillity is somewhat unclear. The United Nations has said that no country can claim sovereignty over a celestial body, but it doesn't say anything about individuals. Dennis Hope, the Lunar Embassy's founder, says that both Ronald Reagan and Jimmy Carter own plots, as well as Tom Cruise and Harrison Ford. There may soon be a boom in intergalactic real-estate litigation. If only NASA had thought of that earlier, it wouldn't have needed the cold war.

For 18 years the Lunar Embassy of Rio Vista, Calif., has been offering 1,800-acre plots of land on the Moon and the planets for the low, low price of $15.99 (plus lunar tax).

Reach for the Moon[4]

Whenever I look at the Moon I do not immediately look towards the Sea of Tranquillity where man first landed on it, but towards a nearby region called Taurus-Littrow. It borders the eastern shore of what early observers called the Sea of Serenity. These days we know it isn't a sea, but the unchanging mountains and plains of the Moon do have a serenity and tranquillity of their own. Taurus–Littrow is on a tongue of lunar highland that separates Serenity from Tranquillity. Of all the places where man touched down on the Moon, the general region of *Apollo 17*'s Taurus–Littrow landing site is the easiest to see through a telescope. The exact valley, however, is elusive. It needs high power and still conditions. But when the night air is calm you can zoom in and peer down on the last place where a man walked on the Moon.

His name was Gene Cernan and 25 years ago he and geologist Harrison Schmitt had just explored Taurus–Littrow and ended an era. As he completed his final walk, Cernan grabbed the TV camera and pointed it at his lunar craft's front landing gear. Here was a plaque with words that sounded so final. He removed the cover and read, "Here man completed his first explorations of the Moon, December 1972."

He went on to add: "This is our commemoration that will be here until someone like us, until some of you who are out there, who are the promise of the future, come back to read it again."

With one last look around, he climbed the ladder into the spacecraft to begin his journey home. The Earth was high in the south-western lunar sky.

The race to the Moon was the gamble of a dead president. It had begun 15 years earlier with the launch of *Sputnik 1* by the Soviet Union. It shocked America, and in turn America's fearful reaction shocked President Eisenhower. Yet he never fully realised the importance of space in either military or propaganda terms. Having lost the first rounds in the space race it was left to another man to take up the challenge. President Kennedy made a speech in May 1961 that set the U.S. on course for the Moon: "I believe this nation should commit itself to achieving the goal, before this decade is out, of landing a man on the Moon and returning him safely to the Earth."

Kennedy made a speech in May 1961 that set the U.S. on course for the Moon: "I believe this nation should commit itself to achieving the goal, before this decade is out, of landing a man on the Moon and returning him safely to the Earth."

---

4. Article by David Whitehouse. From *New Statesman* p26-28 Dec. 12 '97.Copyright © 1997 *New Statesman*. Reprinted with permission. Dr David Whitehouse is the BBC's science correspondent.

Some believe the race to the Moon prevented military confrontation between the superpowers, that it was the moral alternative to war.

On 21 December 1968 man finally left the cradle of the Earth. In orbit around the Earth, the third stage of the giant Saturn 5 rocket was reignited, and three men broke free of Earth's gravity for the first time and began the voyage towards another world.

Astronauts Frank Borman, Jim Lovell and Bill Anders were the first men truly to see the Earth as a planet. Each hour as they looked behind them it dwindled a little more, being swallowed by the great cosmic dark. Soon they could see beneath them a landscape like the aftermath of the final battle, and crossing the lunar limb they saw the radiant blue and white Earth hanging over the cold, grey lunar horizon. In a sense, we had gone to the Moon to discover the Earth, beautiful, fragile and small. Man's only home.

The first landing on the Moon took place in July 1969. Almost by accident Neil Armstrong and Buzz Aldrin were chosen for that historic mission. In a roster of crew assignments drawn up before the date of the landing was decided, they were given *Apollo 11*. By the time *Apollo 8* reached the Moon it seemed likely that *Apollo 11* would be the first to attempt a landing. Armstrong believed the chances of a successful touchdown were 50-50 and before launch, during a dinner with the head of the American space agency NASA, the crew of *Apollo 11* was given a promise. NASA's head, Tom Paine, said that if they failed to land, the next mission would be theirs, and the next, until they succeeded.

Back on Earth, quietly one morning, someone placed a handwritten note next to the eternal flame on Kennedy's grave. It read: "Mr. President, the Eagle has landed."

No one who witnessed those translucent black-and-white images from the Sea of Tranquillity will forget them. Here was man on another world, hopping around in the ultimate desert, silent and stark, on a tiny world where the horizon was only one and a half miles away. Back on Earth, quietly one morning, someone placed a handwritten note next to the eternal flame on Kennedy's grave. It read: "Mr. President, the Eagle has landed."

But it was clear that things were changing on Earth. At an Apollo party President Nixon said, "Here's to the Apollo programme. It's all over."

In a way he was right. Apollo had achieved its objective and there wasn't much else for it to do. You wouldn't ask Lindbergh to fly the Atlantic again, it was said. NASA had enough hardware for nine more landings, leading up to a grand finale: *Apollo 20*, a touchdown in the dramatic Copernicus crater.

It looked easier than it was. Months later *Apollo 12* descended into the Ocean of Storms and when Dick Conrad stood on the surface he looked into the middle distance. There, 600 feet away, on the slope of an ancient crater was *Surveyor 3*, which had landed two and a half years earlier. By the time *Apollo 12* returned, *Apollo 20* had been cancelled and *18* and *19* looked uncertain.

Apollo 13 limped back to Earth after an explosion destroyed all hope of a landing. Until the blast, public interest in the mission had been low. NASA's efforts to keep the crew alive in their clammy capsule for 87 hours, in their thin flight-suits, with the interior temperature dropping to that of a refrigerator, and then seeing them through re-entry, was a tour de force that gripped the world. When the crew returned, however, the White House let it be known space exploration no longer held such a high position in the national list of priorities. Hopes of a manned flight to Mars were gone and the space shuttle and space station were in peril.

The Stars and Stripes had been planted on its barren plains, the Russians were beaten and now there were other things on people's minds.

Apollo was wound down. The programme's final three Moon landings, to some the most exciting of all, were cancelled. For others, the desolate Moon no longer held any fascination. The Stars and Stripes had been planted on its barren plains, the Russians were beaten and now there were other things on people's minds. Nixon was reducing America's global responsibilities as it faced new limits on its resources and will. The "pay any price, bear any burden" attitude of the Kennedy-Johnson years was over. Inflation went unchecked. Paine resigned.

But there were a few last glories. *Apollo 14* landed in the Fra Mauro highlands, touching down on material ejected billions of years ago from the great Imbrium basin 400 miles to the north. With mission 15 the Apollo programme really hit its stride. It landed near a vast chasm in the lunar surface called Hadley Rille. During a moonwalk David Scott saw a small white rock which he immediately recognised as anorthosite—part of the Moon's primordial crust, the so-called Genesis rock, 4.5 billion years old. *Apollo 16* touched down in the central highlands and *Apollo 17* in a steep walled canyon. And then it was over. The cold, grey Moon was alone again. One flight controller back on earth remarked that it must have felt the same when they finished the Pyramids.

That was how, 25 years ago, a great adventure ended, and it is perhaps in the nature of things that we look back, sensing that we did not know what we had until it had gone. Many, even today, haven't truly comprehended the historic nature of our first journeys to another world. It was an undertaking as heroic and as perilous as any made by anyone in any age. No other explorers, no Jason, Marco Polo or Columbus, no Vasco da Gama, no Cook, Amundsen or Hillary matched the voyages to the Moon.

But as well as history what else did Apollo leave us? It inspired a generation to become involved in a grand engineering feat, which fertilised many other areas of American industry. Microelectronics, computers, avionics, advanced materials all benefited from Apollo. New manufacturing processes had to be developed to produce the mighty Saturn 5 rocket, without which the Moon flights would have been impossible. Years later President Bush said that Apollo was the best return on an investment since Leonardo da Vinci bought a sketchpad. Apollo, more than any other space project, showed us where we live in the universe and started a global ecological awareness that lives on. And, not least, Apollo also gave us the Moon.

The Moon was born in 15 minutes. Billions of years ago the hot, volcanic terrain of a primordial Earth, with perhaps the first flickerings of life on its seared surface, was struck by a planetary body the size of Mars. Within seconds this titanic collision squirted molten and vaporised rock into space.

Much of it escaped to become debris orbiting in between the planets, but some remained in orbit around Earth and collapsed into a ball that became our Moon. After that, night-times on Earth would never be the same again.

Being small, the Moon rapidly lost its internal heat and was volcanically active for only a short part of its life. Now, with a surface largely unchanged for aeons, it is like a four and a half billion-year-old book recording the history of our solar system. Thanks to Apollo we know more about the Moon than we do about any other object in space, with the exception of our own planet. We have the testimonies of those who went there—and what tales they are. We have around 2,000 samples from nine sites, 382 kg from six Apollo landings as well as 0.3 kg from three Soviet automated sample return probes.

The samples tell us that the Moon is rich in minerals. The surface has also been examined at ten additional sites by other probes. We have measurements made in orbit by the Apollo spacecraft, we have a partial photographic survey carried out by the Lunar Orbiter probes in the 1960s and a complete multispectral survey performed recently by the Clementine mission, which means we know more about the surface of the Moon than we do about some deep sea areas of the Earth.

Years later President Bush said that Apollo was the best return on an investment since Leonardo da Vinci bought a sketchpad.

Returning to the Moon will be very different from the first time around. It is, however, the only real way to get the public interested in space again. The space shuttle doesn't go anywhere, except round in circles, the space station is worthy but dull and a trip to Mars is just too expensive. To capture the public's imagination it just has to be a trip to the Moon.

The science possible there will be astounding. Just think how it differs from the Earth and how those conditions could be exploited: low gravity, no magnetic field, no atmosphere, no water, high vacuum, low or high temperature depending on whether or not you're in shadow, seismic stability, no radio interference on its far side and total sterility. Astronomers could use it as a unique platform to probe the Universe as well as to collect particles thrown off by the Sun. Its low gravity could enable new materials to be developed and provide new insights into biology. It could provide a site for accurate experiments that need controlled temperatures and stability . . . and perhaps one day it will also be a playground.

Returning to the Moon will be very different from the first time around. It is, however, the only real way to get the public interested in space again.

To build the first Moonbase we need take nothing with us except our ingenuity and perhaps some hydrogen. There in the lunar rocks is everything required for supporting life and many profitable industries as well. Oxygen makes up to half the weight of Moon rocks. It is straightforward to extract, requiring only energy, which could come from small nuclear plants or from solar power. Test rigs have successfully extracted oxygen from simulated lunar soil using a simple and reliable furnace. Automated oxygen factories could even be sent to the Moon in advance of the next explorers. The value of oxygen made on the Moon in life-support systems is obvious. But oxygen can also be used as a rocket fuel and its manufacture from Moon rocks would be the lifeblood of a viable lunar colony.

Because the Moon's gravity is much lower than that of the Earth it requires far less energy to mine and ship rocket fuel from the Moon to Earth's orbit than it does to get it up from the Earth. In fact it requires far less energy to send rocket fuel from the Moon to Mars than it does to get it from the Earth's surface into Earth's orbit. If we ever explore space the Moon has a golden economic future providing rocket fuel.

The Moon's land area is similar to that of Africa. It's as rich in minerals and, in the future, it will make fortunes for as many people as Africa once did. It is only three days away from Earth and in 100 years' time the LunOx Corporation could be the biggest company on or off the Earth. Corporations in Japan have studied the economic viability of extracting minerals from the lunar surface and have supported a series of small space probes to reconnoitre the Moon in the first years of the next century. They know that whoever controls the Moon's oxygen controls the solar system.

V. New Technologies and Discoveries

Editor's Introduction

Just a few years ago, things like electronic mail and the Internet were known and used by a few universities and no one else. Today, such technology is now almost as basic to communications as the telephone. Not long ago, the mapping of a person's DNA sequence or the ability to clone an animal was science fiction. Today, researchers are discovering more about genetics and what makes up living tissue almost daily. As we learn more about ourselves and our planet and find new and easier ways to disseminate and apply that information, we are also learning more about our solar system and universe and are devising new and better ways to study them.

Arguably, the single most important space-based discovery of the decade was what has come to be known as the Mars Rock. In it, NASA scientists claim to have found traces of microscopic biological material—the first real evidence of extraterrestrial life. Further analysis of the Jupiter moon Europa by the space probe *Galileo* has uncovered intriguing new evidence of an ocean beneath the moon's icy surface. Most scientists now believe that Europa is heated by internal volcanic activity and may very well support a thriving ecosystem in its dark ocean. Another important revelation of the last couple of years was the discovery of planets in distant galaxies, which now bolsters the almost universally accepted theory that ours is not the only solar system in which planets exist.

As space travel enters its second century, the aerospace industry is working to develop new modes of transportation and more efficient engines. The first and most ambitious of these inventions is the ion engine aboard the unmanned probe *Deep Space 1*. Such an engine is 10 times more efficient than conventional rockets and can be fired for months or years as it continually builds acceleration. The probe also has a number of new technologies that NASA is testing, including a type of software with artificial intelligence that enables the probe to essentially run its own mission. Lockheed Martin is currently developing a new type of shuttle for NASA, called the X-33, which will have a number of improvements over the current shuttle: a single-stage-to-orbit capability, a greater flight capacity per year, and an overall lower cost per flight. As NASA looks to leave the transportation of satellites and supplies to the space station to the private sector, such a vehicle, as well as the new unmanned rocket planes being proposed by Kistler Aerospace, will be in keeping with the space agency's new ideal: "better, faster cheaper." These prototype vehicles are just that. They have the ability to put satellites in orbit cheaper and more frequently than conventional rockets, while at the same time being reusable.

The purpose of Section V is to provide the reader with some detail about the major discoveries of the past few years, as well as the recent technological breakthroughs in space travel. This section is divided into two parts: the first covers new technologies that are being developed for the space program, the second covers recent discoveries. In the *New York Times* article, "Ion Propulsion of Science Fiction Comes to Life on

New Spacecraft," Warren Leary fleshes out the Deep Space 1 mission, its state-of-the-art ion engine, and the other new technologies it has on board. In the article, Dr. Marc D. Rayman of NASA's Jet Propulsion Laboratory notes that *Deep Space 1*'s technology is completely revolutionary. "It's like having one's car find its own way from Los Angeles to Washington, D.C., and park itself within one foot of its destination, all the while getting 300 miles to the gallon of fuel," he says, adding: "This has the potential to change everything." The next two articles look at new ways to haul satellites and space station supplies into orbit. The first, "Kill the Shuttle? RLV Debate Heats Up," by Joseph C. Anselmo for *Aviation Week and Space Technology*, focuses on the growing debate about whether or not NASA should retire the space shuttle in favor of a newer single-stage vehicle. The next, "Rocket Planes," by Bill Sweetman for *Popular Science*, looks at several designs of unmanned rocket planes currently being marketed to NASA, which will be able to lift satellites and other technology into orbit. The next four articles look at some of the recent discoveries in space. In the *New York Times* article "After Mars Rock, A Revived Hunt for Other-Worldly Organisms . . . ," William J. Broad compares the fossilized evidence found in the Mars Rock and the new evidence of an ocean on Europa to analogous conditions on Earth in which life might have evolve. The next article, "An Ocean of Seltzer," by Kathy A. Svitil, was published in *Discover* and reveals specific information relayed from the Galileo probe orbiting Europa, whose analysis indicates a carbonated ocean beneath the surface. In another *Discover* article, "Impossible Planets," Sam Flamsteed reports on new Jupiter-sized planets orbiting distant stars and the scientific community's quest to figure out how they came into being. In the final article, "JPL's Interferometry Mission to Hunt for New Planets," published in *Aviation Week and Space Technology*, Michael Mecham writes about future endeavors by NASA's Jet Propulsion Laboratory to hunt new planets in distant galaxies through a series of experimental probes and telescopes.

When one looks at the vast array of recent breakthroughs, both inside and outside the space industry, it appears that whatever human beings can imagine, they can eventually create. When one looks at the amount of unprecedented discoveries over the last few years, one cannot help but feel that the universe has become a little smaller, as more and more evidence suggests that the unique conditions that have facilitated life on this planet are not so unique after all.

Ion Propulsion of Science Fiction Comes to Life on New Spacecraft[1]

Washington—Combining technology with fantasy, the United States is about to send an experimental probe into space to test some of the favorite concepts of science fiction. The robot craft not only will be driven by an advanced ion propulsion engine, it will also take care of itself, navigating through space on its own and occasionally calling home to let people on Earth know how it is doing.

The spacecraft, called *Deep Space 1*, is to be a testbed for new technologies that may be incorporated into the next generation of probes sent to explore the solar system and beyond in the next century. Engineers at the National Aeronautics and Space Administration say the mission, scheduled to be launched on Oct. 25, is one of the first designed to send a spacecraft away from the Earth with technology testing, and not science, as the main objective.

Dr. Wesley Huntress, NASA's associate administrator for space science, said *Deep Space 1* was part of a new program to examine and validate unusual technologies that could be used to produce more efficient, cheaper scientific spacecraft. Those planning science missions are sometimes reluctant to risk trying promising, but untested, new technologies or approaches, he said.

High-risk, but relatively low-cost missions like *Deep Space 1* should give planners more confidence in trying some of these new ideas in their projects, Dr. Huntress said. Appropriately, this harbinger of the future space program comes on the heels of NASA's 40th anniversary, which was celebrated last week.

Deep Space 1, a 1,000-pound craft crammed with miniature instruments and devices, is to test 12 innovative technologies in its first few months in space. Among the technologies aboard are an autonomous navigation system; a miniature integrated camera and imaging spectrometer; an integrated suite of space physics instruments for studying charged particles, or plasma, flowing through space; advanced software with artificial intelligence that lets the spacecraft help run its own mission as situations change, and a variety of new, low-power electronics.

The robot craft not only will be driven by an advanced ion propulsion engine, it will also take care of itself, navigating through space on its own and occasionally calling home to let people on Earth know how it is doing.

---

1. Article by Warren E. Leary. From the *New York Times* F p1 Oct 6, 1998. Copyright © 1998 the New York Times Company. Reprinted with permission.

If all goes well, *Deep Space 1*'s primary mission will culminate in July when the craft makes a very close flyby of a small asteroid, 1992 KD. Later, if the spacecraft is healthy and NASA chooses to extend the mission, the probe will fly on to visit a dying comet known as Wilson-Harrington in January 2001. After that encounter, the craft could visit the comet Borrelly, one of the most active to regularly visit the inner solar system, in September 2001.

The mission is expected to cost $152 million, including $42 million for the conventional rocket used to fire the spacecraft away from Earth, project officials said.

But most of the attention being paid to *Deep Space 1* is focusing on its solar-electric propulsion system, a variation on the concept of ion drive that has been a mainstay of science fiction for decades. While ion propulsion units have been tested in space before, and small ion engines have recently been used to stabilize communications satellites in high Earth orbits, the new mission will mark the first time that this type of engine will be used as the primary drive system for a spacecraft.

While ion propulsion units have been tested in space before . . . the new mission will mark the first time that this type of engine will be used as the primary drive system for a spacecraft.

"I first heard of ion propulsion in 1968 during a *Star Trek* episode," said Dr. Marc D. Rayman of NASA's Jet Propulsion Laboratory in Pasadena, Calif., the chief mission engineer for *Deep Space 1*. An ion engine uses fuel 10 times more efficiently than conventional rockets, he said, but it produces an almost imperceptible amount of thrust. This type of engine works by firing continuously for months or years, gradually adding momentum to a spacecraft in the vacuum of space, instead of burning all of its fuel in a few minutes at high thrust, as conventional rockets do.

"Ion propulsion is acceleration with patience," Dr. Rayman said.

Deep Space 1 will be sent on its way from Cape Canaveral, Fla., boosted by a three-stage Delta rocket. Between two and four weeks after launching, engineers will begin testing the ion drive, running it for a 10-day period that will be interrupted every other day for navigation tests. After the shakedown period and a couple of weeks to evaluate the results, the engine will be switched on for long-term thrusting that may continue for several months.

John F. Stocky, manager of propulsion technology at the Pasadena laboratory, which is managed by the California Institute of Technology, said the craft would have 186 pounds of xenon gas aboard to fuel the engine, ample supplies for the primary and extended missions. The xenon, a heavy, inert gas, will not be burned like conventional rocket fuel, but charged electrically to create high-speed particles that leave the engine so fast that they produce thrust.

"We are trying to turn into fact today what was science fiction yesterday," Stocky said.

When the ion engine is running, he said, xenon is injected into a chamber and bombarded with high-speed, negatively charged electrons produced by heating a barium-calcium-aluminate material carried in tanks near the engine. These electrons, guided by magnetic rings around the chamber, hit the xenon atoms, knocking away one of the 54 electrons orbiting each atom's nucleus. The resulting atoms, with a net positive charge, are known as ions.

At the end of the chamber, which opens into space, are two metal grids made of molybdenum that are charged positively and negatively, respectively. The charges in the grids create an electrostatic pull on the xenon ions, accelerating them out of the engine and into space at more than 62,000 miles per hour, producing 3,000 tiny beams of thrust.

Stocky said it was important to keep the engine chamber and the rest of the spacecraft electrically neutral by keeping the ions moving away. If a charge builds up in the spacecraft from the escaping plume, ions leaving the chamber might come back and negate any thrust. So an electrode at the rear of the engine shoots electrons into the glowing blue ion stream leaving the grids, replacing missing electrons and neutralizing the electrical charge of the escaping xenon.

The engine on *Deep Space 1* should produce about one-250th of a pound of thrust, Stocky said, comparable to the force exerted by a single sheet of typing paper resting on the palm of a person's hand. But this steady thrust increases the speed of the spacecraft by about 30 feet per second each day, enough to increase *Deep Space 1*'s speed over the life of the primary mission by 8,000 m.p.h., engineers said.

At times, the ion engine will be turned off and the spacecraft will coast, engineers said. The spacecraft will stop the engine once a week to take its bearings and send routine signals to Earth, they said, and it also will remain off for periods of months in an extended mission when extra energy is not needed as the spacecraft shapes its orbit around the Sun.

To provide the electrical power to run the engine, *Deep Space 1* will use a new type of solar power array never before flown in space that concentrates and focuses sunlight onto 3,600 electricity-producing solar cells. The craft will have a pair of solar wings that expand to 38.6 feet across when fully extended and these arrays will produce 15 to 20 percent more power than similar devices the same size.

Atop the solar cells sit 720 cylindrical lenses made of silicone. The lenses, which look like glass cylinders cut down the middle, gather incoming sunlight on the rounded outer sur-

"Forget the Dilithium Crystals, Bring On the Ions"

Straight from the pages of sci-fi novels, Deep Space 1's primary thrust will be provided by NASA's first operational ion engine. The new propulsion system uses just one-tenth the amount of fuel of a chemical rocket, making exploration of the outer solar system more practical.

How It Works

1. Negatively charged electrons are generated by a solar-powered cathode, which sits at the edge of a chamber filled with xenon, a heavy inert gas.

2. As the electrons are guided by magnetic fields toward the positively charged walls of the chamber, they strike xenon atoms, knocking off the atoms' electrons. The result: positively charged atoms called ions.

3. As these ions drift toward the open end of the engine, they are drawn to the neutral vacuum of space. Two charged grids focus the exiting ions into a high-velocity beam, producing thrust.

4. To avoid building up a negative charge in the spacecraft, a second cathode emits an electron stream into the ion beam, restoring DS1 to its neutral state.

(Source: NASA / Jet Propulsion Laboratory)

faces and focus it down to the solar cells below the flat sides of the half-cylinders. Development of the arrays is co-sponsored by the Pentagon's Ballistic Missile Defense Organization.

Previous unmanned space missions have required ground controllers to constantly monitor conditions on the spacecraft and provide navigational information to keep them on course. Rayman said *Deep Space 1* would require fewer than a dozen controllers on Earth, instead of the hundreds needed for some previous missions, because it would keep track of itself.

About once a week, the spacecraft will take pictures of stars with asteroids in the background and use this information to navigate itself, including having the ability to adjust its course and decide how close it will come to the asteroid and comets it targets.

"Using stars and asteroids as reference points, the spacecraft takes pictures of the sky and infers where it is and where it wants to go," Rayman said.

Instead of constantly sending information on its condition to Earth, the craft will occasionally send special signal tones that signify different states of well-being. The tones range from "everything is operating acceptably" to "there is a problem that may require help if it worsens" to "need urgent assistance from the ground."

All the technologies of *Deep Space 1* combine to point to an entirely new way of designing and operating robotic scientific missions away from Earth, Rayman said.

"It's like having one's car find its own way from Los Angeles to Washington, D.C., and park itself within one foot of its destination, all the while getting 300 miles to the gallon of fuel," he said, "This has the potential to change everything."

Kill the Shuttle? RLV Debate Heats Up[2]

A report prepared for NASA by an outside consultant recommends dropping the space shuttle fleet if a new RLV such as Lockheed Martin's *VentureStar* emerges from NASA; X-33 single-stage-to-orbit technology program.

The report by Hawthorne, Krauss & Associates of Boston concluded that switching to a *VentureStar*–type vehicle for station servicing and other missions could cut NASA's annual launch costs by about two-thirds, to $825 million. That would free up $1.6 billion a year for other space endeavors.

But shuttle supporters quickly dismissed the report, saying its conclusions were based on flawed assumptions and limited input from industry. They predicted NASA's required investment in a new RLV such as *VentureStar* could far exceed the report's $1.75 billion figure.

With such debates likely to grow louder next year as the X-33 moves toward its first flight, NASA is keeping its options open. The agency is laying the groundwork to enable it to utilize a new RLV if one emerges and a decision is made to scrap the shuttle fleet.

One NASA study is looking at phasing out shuttle flights to the station in 2004–05, replacing the five annual shuttle servicing missions with 10 flights of the smaller *VentureStar*. Six of the RLV flights would be unmanned cargo missions, and four would be dedicated solely to rotating station crews, according to Bill Cirillo, a senior engineer at Langley Research Center, who is managing the study.

VentureStar's commercially driven design is tailored to carry unmanned payloads into low-Earth orbit for $1,000/lb. and quickly return to Earth, landing like a shuttle. The windowless vehicles would carry their cargo in standardized containers that could be changed out easily. Cirillo said NASA's study is looking at what modifications would be needed to put *VentureStar* in compliance with new "human rating requirements" issued by the Johnson Space Center in June. The requirements mandate, among other things, a reliability level exceeding 99 percent, a crew escape system, and the ability for the crew to take manual control of the vehicle during all phases of flight.

One NASA study is looking at phasing out shuttle flights to the station in 2004–05, replacing the five annual shuttle servicing missions with 10 flights of the smaller Venture-Star.

2. Article by Joseph C. Anselmo from *Aviation Week and Space Technology* p 67-69 Nov. 30, 1998. Copyright © 1998 *Aviation Week and Space Technology*. Reprinted with permission.

NASA is considering options to either add windows for crew flights or use a synthetic external vision system. "At this point, we don't see a stick-and-rudder kind of function" for the crew, said Gene Austin, NASA's X-33 program manager. "We see more of a computer interface."

Austin said if a new RLV becomes a reality, the goal would be to phase out the shuttle, turning space transportation over to the private sector so NASA can concentrate on exploration. While VentureStar would be much less versatile, key shuttle activities such as scientific research will have already migrated to the new station, he said.

NASA sees a need for a minimum of 13 RLV flights a year—the 10 station flights, plus at least three launches of unmanned spacecraft. That would account for just under one-third of the 40 annual VentureStar flights currently envisioned.

The Hawthorne report estimated NASA would need to invest $1.75 billion in a human-rated VentureStar. The agency would spend $825 million a year, $75 million for each of the 10 station flights and $25 million for the three nonstation flights. By contrast, the report found that investing in unspecified shuttle upgrades would cost $5.2 billion and shave just $60 million off the $300 million cost of each mission.

The study said a new RLV would not only generate cost savings, but significantly enhance the potential of the space station, "whose commercial success is heavily dependent upon reduced access costs."

Russ Turner, president and CEO of United Space Alliance, the Lockheed Martin–Boeing venture that operates the shuttle for NASA, charged that the report's conclusions were based on "many specific errors of fact."

He said the report's authors had consulted with United Space Alliance, but "they only cited one member of industry, VentureStar."

Turner said the report failed to address Lockheed Martin's need for loan guarantees to develop VentureStar and incorrectly stated that United Space Alliance has shown no interest in funding major shuttle upgrades, when in fact the company has already invested $80 million in shuttle improvements.

He said the document also doesn't take into account how the unique shuttle/station capabilities would be replaced after the station is gone. The station is being designed with a 10-year lifespan, though NASA officials say it may ultimately stay in orbit significantly longer.

[NASA needs] a minimum of 13 RLV flights a year—the 10 station flights, plus at least three launches of unmanned spacecraft. That would account for just under one-third of the 40 annual VentureStar flights currently envisioned.

Similar criticism came from officials at Boeing, who are pitching costly upgrades for the shuttle, such as a reusable liquid flyback booster.

Rick Stephens, vice president and general manager at Boeing Reusable Space Systems, said a space transportation architecture study Boeing and four other companies are performing for NASA will provide a much more in-depth analysis of the tradeoffs between keeping the shuttle or proceeding with a new RLV. The study is expected to be completed in January.

Stephens said the current shuttle fleet has 80 percent of its flight life remaining and could operate beyond 2030. "The whole question and debate on where the nation should invest its resources . . . is going to go on for a number of months." he predicted.

Finance is just one part of the equation, however. The technical success or failure or the X-33 program will drive the timetable on development of a new RLV.

"Technical problems with the rocket engines recently forced NASA and Lockheed Martin to delay the first flight of the suborbital test vehicle by five months, to December 1999. Lockheed Martin officials have said they hope to make a decision on whether to proceed with *VentureStar* shortly after the X-33 flight test program is completed.

Rocket Planes[3]

Reusable launch vehicles could revolutionize how we host satellites into orbit, deliver packages, and even travel the globe.

It's all part of a revolution in access to space: a change from big, expensive, one-shot rockets funded by government agencies to smaller, less costly, reusable vehicles designed and built by private industry.

Above Edwards Air Force Base, a C-141A military cargo plane tows a veteran F-106 fighter through the sky on a 700-foot cable. High over the California desert, the cargo jet releases the F-106 for a glide back to the base. The test is a milestone in the development of the *Eclipse Astroliner*, a reusable launch vehicle designed to deploy communications satellites in space.

The *Astroliner* is an airplane-like vehicle that will be towed up to 40,000 feet by a Boeing 747, whereupon it will fire its rocket engines and zoom to 75 miles above Earth. After popping its nose open to discharge a satellite, the *Astroliner* will glide home for a runway landing.

Eclipse Astroliner is only one of several reusable launch vehicles, or RLVs, under development. In Sacramento, a team of ex-NASA engineers is designing a reusable rocket with engines that were built in the 1960s to put a Russian cosmonaut on the moon. On the western edge of California's Van Nuys airport, a 1950s-vintage test facility powered by a battery of jet engines blasts air through an experimental scramjet engine at eight times the speed of sound. Another team is working on a rocket that lands like a helicopter. Among the backers of RLV development efforts are aerospace legend Burt Rutan, former astronauts Buzz Aldrin and Pete Conrad, and bestselling author Tom Clancy.

It's all part of a revolution in access to space: a change from big, expensive, one-shot rockets funded by government agencies to smaller, less costly, reusable vehicles designed and built by private industry.

It doesn't take a rocket scientist to see something wrong with how we launch satellites. Today's throwaway launch vehicles are direct descendants of the first intercontinental ballistic missiles. Using these rockets to launch satellites into low Earth orbit (LEO) costs about $10,000 per pound, and the cost hasn't dropped much in recent years.

---

3. Article by Bill Sweetman. From *Popular Science* p 40–45 Feb. 1998. Copyright © 1998 the L.A. Times Syndicate. Reprinted with permission.

This is bad news for the industry consortiums that want to put more zip into worldwide telecommunications with constellations of LEO satellites. The cost of launching these satellites, and replacing them as they age or break down, threatens to ruin the otherwise promising economics of global phone, data, and video services.

Lower launch costs are the aim of the NASA/Lockheed Martin X-33 prototype and its operational follow-on, *VentureStar*. But *VentureStar* won't fly until 2004 at the earliest, and its first task may well be to replace the shuttle. Designed to launch sections of the International Space Station, *VentureStar* is also awkwardly big for the small LEO satellites.

That makes the satellite market a plump target for a group of small start-up companies that plan to build new launch vehicles with funding from private investors. So far, none of the companies has a working prototype, nor more than a few dozen employees, and it's likely that some of them will run out of money before they get the job done. But it's also quite possible that one or more of these companies will succeed in building the world's first privately funded reusable spacecraft. Such a craft could not only launch small satellites and scientific research platforms, it might also deliver packages or even passengers across the globe much more quickly than can an airplane.

In theory, many different types of reusable launch vehicles can be built. They can be piloted or unmanned. They can take off like an airplane or a rocket. They can blast all the way from the ground to space in one giant step, or they can use two stages to get there. The five start-up RLV companies leading the pack all offer different solutions.

The airborne tow tests over Edwards are designed to prove a key element of Kelly Space & Technology's concept for a winged space plane. Michael Kelly is a former TRW executive who started his company in San Bernadino, California, after TRW canceled the launch vehicle project he was heading. From the outset, Kelly says, he wanted a launcher that could fly from any airport, but he knew that a piggybacking vehicle would be limited in size by the payload of the aircraft carrying it. "Then we decided to tow the thing, and we got more relief than we dreamed possible," he recalls. World War II tests of giant troop-carrying gliders show that a 747 could pull a 500,000-pound airplane aloft.

Kelly has an $89 million contract with Motorola to launch 20 of the company's Iridium LEO satellites on the *Eclipse Astroliner*, starting in July. Meanwhile, the company is building a smaller prototype, called the *Eclipse Sprint*, scheduled to fly in 1999. About the same size as the F-106, the *Sprint* will be a testbed for the *Astroliner*'s flight controls, its aero-dynamics, and its structural materials—aluminum alloy protected by thermal blankets and tiles. Development of the full-scale vehicle will cost only $150 million, thanks in part to the use of modified parts from other aircraft.

The *Astroliner* will be built like a conventional airplane, flown by two pilots, and powered by an off-the-shelf rocket motor. It will not go all the way to orbit, but will reach a speed of Mach 6.5 and an altitude of 75 miles before its nose pops open and it ejects a satellite, attached to a simple solid-rocket upper stage.

Kistler Aerospace Corp. plans to beat the *Astroliner* into space, making its first orbital flight late this year. The company's effort began in 1994, when Seattle inventor and entrepreneur Walter Kistler invited financier and developer Robert Wang to join a space launch project. Wang agreed, on the condition that Kistler also take on a manager with experience in an operational space program. The company recruited George Mueller, former manager of the Apollo and Saturn programs at NASA.

Mueller brought in a team of ex-NASA colleagues, who devised a design that he calls "the simplest possible implementation of a transportation system" for lofting LEO satellites. The plan is to build an unmanned two-stage rocket, powered by Russian NK-33 and NK-43 engines burning liquid oxygen and kerosene.

The Kistler K-1 rocket stands 115 feet high, weighs just over 800,000 pounds, and can put about 10,000 pounds into orbit. It looks about as sexy as a feed silo, and its engines—built for the giant N-1 moon shot booster—have been in a Russian warehouse for a quarter of a century. But Kistler is in the lead among the new entrants, with a contract from Loral Space & Communications to launch 10 of the satellites in Loral's proposed Globalstar mobile phone network.

Kistler and other contenders are using Russian rockets because they are far better than anything made in the United States, which has developed only one new liquid fueled rocket motor since the Apollo program. The Russians have developed dozens, inventing ways to operate at higher pressure, and applying advanced materials to make their motors more durable.

Kistler plans to fly K-1s from the Nevada test site and Woomera, Australia; the latter will serve the Asian market. The company does not intend to sell its vehicles or its technology, but to set itself up as a complete launch service, at $17 million a flight. That translates to a satellite-delivery cost of about $4,800 a pound, less than half the current cost. The plan is to build five vehicles, each capable of making 100 flights. So far, the company has secured about $100 million of the $500 million needed to complete the project.

Pioneer Rocketplane's *Pathfinder*, like Kelly's *Eclipse Astroliner*, has a pilot and needs a "kick stage" to orbit a satellite. The *Pathfinder*, which looks like a small space shuttle, takes off using two Pratt & Whitney F100 turbofan engines and climbs to about 30,000 feet. There it receives liquid oxygen from a tanker aircraft, through a vacuum-jacketed hose. Only then does it ignite its Russian RD-120 rockets and fly above the atmosphere to release the upper stage carrying the satellite.

Like towing, midair refueling boosts performance because the vehicle starts with full tanks at 500 mph and 30,000 feet. Also, if the rocket does not have to operate in the dense lower atmosphere, it can have a larger diameter nozzle that is more efficient at high altitude.

Pioneer's team, which includes experimental aircraft designer Burt Rutan and man-on-the-Moon Buzz Aldrin, has won a $2 million design contract under NASA's Bantam program. NASA is looking for a vehicle to launch small payloads (about 400 pounds) for $1.5 million per flight, and may fund the *Pathfinder* prototype if it proves competitive.

The Pathfinder is smaller than the *Eclipse Astroliner*, because its size is limited by the amount of liquid oxygen that an L-1011 tanker can carry. The space plane weighs 220,000 pounds when it disconnects from the tanker, and can orbit a maximum of 5,000 pounds. The airframe is made of aluminum, covered with thermal-barrier tiles like those used on the B-2 bomber's jet exhausts.

Rotary Rocket—a team of entrepreneurs backed by private investors, including author Tom Clancy—has developed another unique approach to space launch. The *Roton* is a single-stage vehicle that takes off like a conventional rocket, powered by an engine that burns kerosene and liquid oxygen. The engine has an advanced aerospike nozzle, which resembles a traditional bell nozzle turned inside out.

Built into the *Roton*'s nose is an unpowered rotor with telescoping blades that deploy and spin up as the vehicle approaches its landing spot. In the seconds before touchdown, the rotor blades pitch up, trading rotational momentum for lift, and the vehicle lands slowly like a helicopter.

Constructed from graphite composite materials, the 52-foot high *Roton C* will weigh 325,000 pounds at launch--less than half the mass of the Kistler K-l. Its designers, among them Burt Rutan, claim the *Roton* will be able to launch payloads for as little as $1,000 per pound and will be capable of returning with a fully loaded cargo bay. Test launches of the *Roton* are due to start in 1999.

Kelly, Kistler, Pioneer, and Roton publicize their programs on the Internet and in glossy brochures. Not so for Space Access. Based in Lancaster, California, a few miles from the Lockheed Martin Skunk Works, Space Access is the closest thing to a "black" military program that you'll find in the commercial world. Subcontractors and potential customers are bound by strict non-disclosure agreements. Even suppliers don't know all of the details of the project.

Space Access has about 50 employees—a mix, says President Steve Wurst, of veterans who worked on high speed programs such as the X-15; and young, computer-savvy engineers. Wurst says Space Access has not publicized its activities because enough investors are already on board to fund the project, and the company wants to maintain its technical lead over the competition

The Space Access launcher is a two-stage vehicle, with an airplane-like first stage that takes off and lands on a runway. The key to the launch system, says Wurst, is a new engine. Elements of the engine have been tested by subcontractor Kaiser Marquardt, a company with unmatched experience in high-speed, air-breathing propulsion. The Space Access scramjet engine is expected to be seven times as efficient as a rocket, eliminating the need for a tow or inflight refueling.

Although the first company to get off the ground will have an advantage, reliability and safety will also be necessary ingredients for success.

Although the first company to get off the ground will have an advantage, reliability and safety will also be necessary ingredients for success. Customers don't merely want a cheaper item, but a launch service that can offer "aircraft-like operations." The space shuttle takes weeks of work between flights, and *VentureStar* is intended to turn around in seven days. Airplanes turn around in hours or their operators go bankrupt, because an expensive vehicle that flies infrequently and makes only a few hundred trips in its lifetime can't afford to offer low fares.

If space planes and reusable rockets do achieve aircraft-like operations, they may be just as useful for delivering packages here on Earth as in space. Federal Express founder Fred Smith, patron saint of procrastinators, has made no secret of his goal to provide next-day service between any two points on the globe—which means that the packages need to go supersonic. More than a decade ago, Smith offered to put FedEx money into demonstrating a hypersonic cargo plane. It may be a better commercial prospect than a supersonic airliner. Whether you count them by the pound or by the cubic foot, express packages bring in more revenue than passengers, and they don't whine if the Moët & Chandon is the wrong vintage.

By a neat technical coincidence, a hypersonic freighter with a 10-ton trans-Pacific payload is about the same size, and needs the same technology, as the first stage of a two-stage medium-lift launch vehicle. So the development cost can be shared between the space launch and cargo markets. However, Kistler and Rotary Rocket may have trouble persuading foreign governments to accept the idea of unpowered, pilotless, wingless vehicles plummeting toward their countries at hypersonic speed, packing enough kinetic energy to take out a city block.

The U.S. Air Force, meanwhile, wants a military space plane for satellite defense, and to perform reconnaissance and precision strikes against terrestrial targets. Although the White House vetoed the space plane project in October, the operational requirement still stands—and "aircraft-like operations" is the watchword for this project as well.

Further into the future, some companies that are developing reusable spacecraft anticipate taking paying passengers into space. With so many possible market opportunities for a reusable launch vehicle—satellite launches, package delivery, military missions, and even space tourism—the development of a commercial reusable spacecraft seems very much inevitable. Still, the first step is a big one for the small companies competing in this space race.

Paul Czsyz, a former McDonnell Douglas engineer and an expert in hypersonics, compares today's oneway rocket technology to Conestoga Wagons. These wagons carried people and their goods from St. Louis to the West, but making the trip back was nearly impossible. Settlement began with the wagons, but industry could not grow until speculators bet everything on a railroad that made round trips possible. The first satellite lofted by a fully reusable, privately funded launcher could be as important as the driving of the last spike of the transcontinental railroad.

After Mars Rock, A Revived Hunt for Otherworldly Organisms; Jupiter's Moon Europa Could Be Habitat for Life[4]

Europa, a moon of Jupiter, is completely enveloped by water, either frozen or liquid, believed to be as deep as 60 miles in some places. In contrast, the Earth's seas reach down barely seven miles at most.

Last week's release of clues that Mars may once have harbored primitive microbes has lent new energy and excitement to a call for close exploration not only of the red planet itself, but of Europa, a moon of Jupiter that some scientists had been independently and quietly examining as a possible home to alien life.

The common denominator of both worlds is water, a prerequisite for life, at least in this part of the universe.

Today, Mars is largely a red desert strewn with rocks and many indications that water flowed over its surface billions of years ago, cutting deep channels and filling large lakes. Mars offers no clear signs that liquid water now runs on its surface; its icy polar caps are composed primarily of carbon dioxide.

But Europa, a moon of Jupiter, is completely enveloped by water, either frozen or liquid, believed to be as deep as 60 miles in some places. In contrast, the Earth's seas reach down barely seven miles at most.

The rub for extraterrestrial life on Europa is that the moon's surface is an icy wasteland. But increasingly, scientists suspect that the Jovian satellite has a hot core and that the inner part of its waters makes up a gigantic dark sea that may seethe with alien life forms that have quietly evolved over billions of years.

So great is the biological allure of Europa that even before the announcement about Mars last week, scientists were making plans to hold a meeting to discuss the odds of life arising there and were lobbying for new exploratory missions to the Jovian moon.

Their excitement has redoubled with the news that a Martian meteorite that fell to Earth contains hints of ancient extraterrestrial life in its recesses.

"It's fantastic," Dr. John R. Delaney, an oceanographer at the University of Washington who is helping to plan the Europa conference, said of the new finding. "With Mars, we're talking about fossil evidence. But where you have a

---

4. Article by William J. Broad from the *New York Times* C p1 Aug 13, 1996. Copyright © 1996 The New York Times Company. Reprinted with permission.

live heat source and a liquid body, you have the potential for living organisms today."

Dr. Joseph A. Burns, a senior planetary scientist at Cornell University who has led national panels that set goals for space exploration, said most experts agreed that after Mars, Europa was the most likely candidate in the solar system for nurturing extraterrestrial life or harboring its fossilized remains. It could be the most likely candidate when it comes to organisms living now.

Interest in exploration of the Jovian moon is rapidly growing, Dr. Burns said in an interview.

"Keep in mind that Europa is not that small," he said, noting that its radius is about half that of Mars. That makes it roughly the size of Earth's moon.

For decades, if not centuries, speculation about the possibility of extraterrestrial life in the solar system focused on planetary surfaces and the idea that the preconditions for living organisms include not only water but also an atmosphere and sunlight, which were seen as providing essential energy and a refuge from the icy cold of space.

But a main discovery of the late 20th century is that rich ecosystems have flourished on Earth in complete darkness for billions of years, drawing energy from planetary heat rather than sunlight. Increasingly, the intriguing question is whether Earth is unique in this respect.

On Earth, the ecosystems that exist without sunlight are in the blackness of the deep sea. They were discovered in 1977 off the Galapagos Islands, along a volcanic rift that winds through the depths of the global sea like seams on a baseball. The otherworldly fauna include giant clams and fields of tube worms; the white casings where the worms live are up to 10 feet long.

It turned out that the dark ecosystems are powered by tiny microbes that thrive on chemicals released along the volcanic rift by Earth's inner heat. The microbes play a role analogous to that of plants in sunlit realms.

Genetic clues have suggested that such microbes are the ancestors of the earliest forms of life on Earth. The heat-loving microbes have been found to be widespread, thriving not only in hot springs on the ocean floor but also in active volcanic craters, deep oil reservoirs and deep, wet rock. They tend to flourish in areas of extreme heat and pressure, as in the subterranean world.

In 1992, Dr. Thomas Gold of Cornell University proposed that the microbes might be ubiquitous throughout the upper few miles of Earth's crust, inhabiting the fluid-filled pores, cracks and interstices of rocks while living off Earth's inner

A main discovery of the late 20th century is that rich ecosystems have flourished on Earth in complete darkness for billions of years, drawing energy from planetary heat rather than sunlight.

heat and chemicals. He calculated that the total mass of this hidden biosphere might rival or exceed that of all surface life.

"Such life may be widely disseminated in the universe," Dr. Gold added in his 1992 paper. These kinds of ideas and terrestrial findings have put a new spin on the old question of extraterrestrial life.

For Mars, the emerging exploratory push fueled by the recent meteorite discovery aims to find evidence of tiny microbes that may flourish deep underground, in the wet and temperate parts of the planet's interior, or, if not such creatures, then fossil remains of them from a bygone era.

But such theorizing becomes far more provocative when it comes to Europa because its deep waters are envisioned as perhaps harboring a thriving mass of alien creatures.

Jupiter is the largest planet of the solar system, and its moons are the system's biggest and most mysterious. Europa, with surface temperatures estimated at minus 230 degrees Fahrenheit, has long been considered to be far too cold to support life, its 60 miles of water thought to be frozen solid from top to bottom.

Scientists have come to believe, however, that Jupiter's very strong gravitational pull keeps its inner moons in a state of geological turmoil. For instance, Io, perhaps the most bizarre satellite in the solar system, is clearly covered by seething volcanoes, which constantly shoot plumes of material into space.

Slightly farther out, Europa, discovered by Galileo in 1610, has no visible volcanoes but is still considered to be geologically alive. The brightest of Jupiter's 16 moons, it has an icy surface made of water that is cut by deep fractures and is clearly very young, as suggested by the relative lack of scars left by comet and asteroid impacts. The surface fissures are thought to be evidence of such internal planetary dynamics as ocean currents.

Moreover, scientists now believe that Europa's gravitational interactions with Jupiter, Io and, still farther out, Ganymede, the largest satellite in the solar system, have caused enough internal friction and heating to keep a large part of Europa's water liquid rather than frozen. The surface ice might be up to 5 or 10 miles thick, scientists believe, making the inner liquid ocean quite extensive, perhaps 40 or 50 miles deep.

Moreover, the rocky floor at the bottom of the conjectural sea might seethe with the same kind of hot volcanic vents as in the Earth's seas, which could make such hot vents on Europa a womb for the origin of extraterrestrial life.

"Whether or not there's submarine volcanic activity today on Europa, it surely was happening sometime in the past," said Dr. Steven W. Squyres, a planetary scientist at Cornell University who has written since 1983 about the possibility of life on Europa.

The rising interest in Europa has resulted in plans for a large conference to examine the question of life there. The event is being organized by Dr. Squyres; Dr. Delaney, of the University of Washington, and Dr. Torrence V. Johnson, of NASA's Jet Propulsion Laboratory in Pasadena, Calif. Financed by the ocean branch of the National Science Foundation and the exobiology branch of the National Aeronautics and Space Administration, it is to be held Nov. 12 to 14 at the San Juan Institute in San Juan Capistrano, Calif.

A push is also developing to try to extend the length of the Galileo mission so the spacecraft, now orbiting Jupiter, can focus on Europa in its final days. The mission is scheduled to end in December 1997, but mission planners would like to have the spacecraft swing past Europa for a series of close images and analyses in 1998 and 1999.

The spacecraft has enough fuel for such an endeavor, they say, but it is not known if money would be available to extend the mission. The payoff of such an extension would be the best portrait yet of the moon.

The first human close encounter with Europa came in 1979 when the *Voyager 2* spacecraft passed within 125,000 miles of it and was able to produce startling images of its icy expanse. Today NASA is releasing photographs of Europa taken by the Galileo probe from a distance of about 100,000 miles. In February, *Galileo* is to fly past Europa at a distance of about 400 miles, repeatedly snapping images.

"We'll be looking for evidence of eruptions on the surface," said Dr. Johnson, of the Jet Propulsion Laboratory, who is the chief scientist of the Galileo mission to Jupiter and its moons.

An extended Galileo mission through 1997 and 1998 would enable the probe to sweep close to Europa six to eight more times, greatly increasing the odds of observing the eruption of liquid water into the moon's thin atmosphere, which last year was discovered to be relatively rich in oxygen. Such a mission would look for activity at the fissures on Europa's surface.

Dr. Johnson said the extended mission might cost $10 million a year, compared with the $50 million it now costs annually to operate the Galileo probe.

Plans are also under way to study the possibility of sending a small new spacecraft to Europa early in the next century that would orbit the icy moon and probe below its surface

The first human close encounter with Europa came in 1979 when the Voyager 2 spacecraft passed within 125,000 miles of it and was able to produce startling images of its icy expanse.

with specially designed radar, perhaps settling the question of whether a hidden ocean exists.

The wildest of the scientific dreams includes having a piloted submarine explore that extraterrestrial sea, if it exists, in a hunt for new geology and life.

Of course, that presumes that any aliens living on Europa would allow the arrival of a craft from Earth.

Arthur C. Clarke, the novelist, over the years has carefully followed scientific theories about the nature of Europa and the possibility of life beneath its icy surface.

In his book *2010*, the sequel to *2001: A Space Odyssey*, Mr. Clarke wrote of an extraterrestrial civilization that beamed a single message to earthlings, repeating it over and over: "All these worlds are yours—except Europa. Attempt no landings there."

An Ocean of Seltzer[5]

For about two years the space probe *Galileo* has been gathering ever more evidence that a large ocean lies hidden beneath the frozen, fractured surface of Jupiter's moon Europa. Now the spacecraft has found the most convincing signs yet for a Europan sea, in the form of salt deposits on the moon's icy surface. The salts very likely come from a briny ocean many miles below the ice. The evidence also suggests that Europa's ocean is highly carbonated; pressure from that seltzerlike sea may be responsible for ejecting what appear to be sprays of debris visible on the moon's face.

The telltale signature of the salts was picked up by an instrument on board *Galileo* called the Near Infrared Mapping Spectrometer, which measures the absorption and reflection of various wavelengths of sunlight by Europa's surface. Each material absorbs and reflects different, characteristic wavelengths of light. The wavelengths measured by *Galileo*'s spectrometer—from deep red in the visible spectrum out to longer infrared wavelengths—can thus reveal the chemical makeup of Europa's surface.

When the spectrometer focused on the areas of Europa that appeared in photographs to consist of pure ice, it indeed measured the characteristic spectrum of water ice—no surprise there. But when the instrument examined features like the many dark lines that crisscross the moon's surface, it obtained several distorted spectra that each looked similar to ice. Thomas McCord, a geophysicist at the University of Hawaii who is studying the *Galileo* data, recognized that the spectra were produced by light reflecting off water-impregnated salty minerals, including natron and Epsom salts. Such salts, says McCord, form only in the presence of liquid water.

"This is the first evidence that Europa's ocean is briny," says McCord. The circulating currents of a salty sea would explain the unusual magnetic readings that other *Galileo* instruments have recorded. Those currents may create a magnetic field for the moon. *Galileo* found that the salts have the same composition at different sites scattered about the moon, evidence that the sea is extensive and well mixed.

The composition of the salts, McCord says, gives some hints about the chemistry of Europa's ocean. The presence of sodium carbonates means that carbon dioxide is probably dissolved in the water. On Earth, volcanoes and submarine

The composition of the salts, McCord says, gives some hints about the chemistry of Europa's ocean. The presence of sodium carbonates means that carbon dioxide is probably dissolved in the water.

---

5. Article by Kathy A. Svitil. From *Discover* p34 Sep. 1998. Copyright © 1998 Buena Vista Publishing Group d/b/a. Reprinted with permission.

hydrothermal vents release carbon dioxide, "so it is reasonable to think that carbon dioxide gas would be coming out of Europa's mantle and into the ocean," McCord says. "The ice cap would seal the ocean most of the time, so the carbon dioxide would build up and become like seltzer water." Raise the pressure of that carbon dioxide high enough and it could erupt through the ice to the surface. (This could explain the origin of volcano-like features on Europa.) Once exposed to the near vacuum that exists at the moon's surface, the briny water would quickly evaporate, leaving the salts behind.

As yet, there is no evidence that Europa has hydrothermal vents on its ocean floor. "But if there is an ocean, there has to be heat coming out of the mantle," McCord says, "so there almost have to be hydrothermal vents." There is certainly no reason that hydrothermal vents couldn't exist in oceans on worlds other than our own. In fact, the *Mars Global Surveyor* mission recently found signs of a large deposit of hematite—a mineral often formed by hydrothermal activity—near Mars' equator. If the hematite deposit does indeed mark the site of a large ancient body of hot water, it would be an ideal place to search for past life on Mars.

A briny Europan sea might also contain evidence of past life. But unlike Mars, Europa may still harbor some kind of life. "It is felt that carbon dioxide in the ocean created a fertile environment for the formation of life on Earth," McCord says. Carbonates create a less acidic environment, one more favorable for life. "Here we have, perhaps, a carbon dioxide-rich ocean. It's one more thing that suggests that the ocean could be a habitat for life."

Impossible Planets[6]

Too bad the perfect one-liner had already been used. When the great Columbia University physicist I. I. Rabi was confronted with news of the muon, a wholly unexpected new subatomic particle, he asked in mock horror, "Who ordered *that*?" Astrophysicists reacted pretty much the same way when University of Geneva observers Michel Mayor and Didier Queloz stood up at a conference in October 1995 to announce they'd found something their colleagues had been seeking for decades—a planet orbiting a sunlike star.

The trouble was, nobody had ordered, or even imagined, a planet quite like the object circling 51 Pegasi, a star lying 50 light-years from Earth in the constellation Pegasus. For one thing, it is huge—about half the mass of Jupiter. Yet despite this bulk, it orbits only some 5 million miles from 51 Peg—seven times closer than tiny Mercury orbits our sun—and whips through one orbit in a scant 4.2 days.

To appreciate how bizarre this behavior is, it helps to consider the bigger planets in our solar system—Jupiter, Saturn, Uranus, and Neptune. They are all at least a hundred times farther from the sun than 51 Peg's planet appears to be. And it takes them years—a full dozen years, in the case of Jupiter—to make a single orbit.

Things get worse when you try to explain how 51 Peg's planet came to be. The only observationally grounded theory of planet formation that astronomers have is the one they reverse-engineered from the only planetary system known to exist (until recently, that is): our own solar system. The story starts almost 5 billion years ago with a slowly revolving cloud of collapsing interstellar gas and dust. The more the cloud falls in upon itself, the faster it spins, eventually flattening into a huge rotating disk. Matter continues to fall into the now rapidly spinning core at the disk's center until it becomes so dense and hot that hydrogen begins fusing into helium, releasing light and other types of radiation: the sun is born. Out in the disk, meanwhile, dust particles collide and stick together until they've built themselves into huge, solid, spherical lumps—such as the planets.

Now comes the part that separates the giant planets from the rest. The relentless pressure of the solar wind sweeps the lighter gases—hydrogen, oxygen, and water vapor, among others—away from the inner solar system, leaving behind

---

6. Article by Sam Flamsteed. from *Discover* p78-83 Sept. 1997. Copyright © 1997 Buena Vista Publishing Group d/b/a. Reprinted with permission.

tiny, naked lumps of dust: puny Earth and its neighbors. Much of this gas blows out to the extremities, to Pluto and the comets. But some of it—quite a lot of it, actually—gets caught up in that vast swath of dust in between. This logjam of gas and dust contains enough matter to make planets 10 times bigger than Earth. Once an object of such size forms, it has a gravitational field powerful enough to act like a giant vacuum cleaner, and it sucks in whatever nearby gas is left over. In a mere 10,000 years or so you have Jupiter—a rocky core ten times the size of Earth surrounded by an immense atmosphere 35,000 miles thick.

Everybody was more or less happy with this story until 51 Peg and its weird planet came along. How could a giant planet form so close to a star without being sucked in by gravity? And where did all that dust and gas come from anyway? Right up against the blast of a strong stellar wind, there shouldn't have been any gas left over, that's for sure. And it's hard to imagine that there would be enough dust available to make something half as big as Jupiter out of rock alone.

In short, who ordered that?

A good astro-nomical theorist never lets the complete absence of infor-mation stand in the way of a nice theory and, conversely, never gets thrown when an actual observation arrives to spoil it.

To assume that the theorists are baffled, though, is to underestimate them badly. A good astronomical theorist never lets the complete absence of information stand in the way of a nice theory and, conversely, never gets thrown when an actual observation arrives to spoil it. Lack of data can actually be an advantage. Unhindered by facts, some theorist is bound to have come up with a model that explains even the strangest discovery.

Doug Lin, from the University of California at Santa Cruz, is a case in point. "I was surprised that Mayor and Queloz found the planet, yes," he admits. "But I wasn't surprised that it existed." In fact, Lin had suggested as far back as 1982 that planets like Jupiter could migrate from the outer solar system in toward their parent stars.

Lin's idea was that if the preplanetary disk was massive enough, the growth of a Jupiter would finally stop, not because there wasn't any gas and dust left to suck in but because there wasn't any within reach. In other words, the voracious planet would have vacuumed up a swath around itself, separating the disk into outer and inner sections, with the planet in between. Remember, though, that there's still a lot of gas and dust remaining in these inner and outer sections. This gas and dust has substantial gravity, which tugs at the Jupiter-like planet. Since the outer disk orbits more slowly than the planet, it tends to slow the planet down and make it spiral in toward the star. The inner disk, on the other hand, whirls more quickly, so it tends to speed the planet up and fling it outward.

Who wins this tug of war? The interaction is complicated, but based on earlier work by other astronomers (who were studying not the formation of hypothetical planets but the interaction between the moons of Saturn and that planet's famous rings), Lin figured out that the outer disk almost always wins. The planet moves inexorably inward, plowing through the dust of the inner disk. "Just a few months before the 51 Peg announcement," he says, "I stood up at a conference and said, 'The reason we haven't found any giant planets yet is that they spiral in.' I can't tell a lie, though. I didn't think they'd stop. I thought they'd continue migrating in until they were swallowed. In that sense, I really was surprised."

It took him maybe an hour to get over it. Within a week of the 51 Peg announcement, Lin and fellow astronomers Peter Bodenheimer and Derek Richardson submitted a paper to the journal *Nature* describing not one but two scenarios for getting a migrating Jupiter to stop short of destruction. The first capitalized on a well-known quirk of T Tauri stars—young, hot, not yet fully formed stars that usually have lots of dust around them. As the young star pulls matter into itself from the surrounding disk, it begins to spin faster, for the same reason that a ballerina's pirouette accelerates as she brings her arms closer to her body. T Tauri stars do this, but they don't often spin as fast as astronomers think they should.

"We know," says Lin, "that they have disks around them, but sometimes we see that the central star is rotating slowly. It doesn't make sense." One possible explanation: These stars might have strong magnetic fields that push out on the disk, creating drag. Just as a thick batter will slow down the whirling blades of an egg-beater, the slowly rotating disk keeps the star from rotating too quickly.

Not only would the magnetic field slow the star's rotation, it would also push away the gas and dust, clearing a space on the disk. Once a Jupiter spiraled into that gap, it would suddenly be free of the disk's influence, and it would stop. "This scenario was possible," says Lin, "but I wasn't happy with it because it requires a tooth fairy—the magnetic field. Also, it doesn't account for the fact that some T Tauri stars are fast rotators." So he and his colleagues offered scenario number two, which doesn't depend on any gap. As the giant planet migrates inward, it eventually gets close enough to be affected by the star's rotational energy, which acts to speed the planet up. Now the tug of war becomes even, and the forces are perfectly balanced. The planet becomes stuck right where it is.

Although no one knows, of course, whether this really happens, for the time being Lin's model is at least plausible. "The intuitive leap Doug made is very attractive," admits Alan Boss, a rival theorist at the Carnegie Institution of Washington. "It's sort of inevitable that this would work. I think it's a positive idea." And after all, the peculiar planet at 51 Peg needs explaining somehow.

Or perhaps it doesn't. Last winter, astronomer David Gray of the University of Western Ontario in London announced that the planet does not in fact exist. "It's not there," he says categorically. "I've ruled it out." Gray's claim is based on the way the planet was discovered. Mayor and Queloz never spotted the planet circling 51 Peg directly—even an absurdly large planet is impossible to see when it's 50 light-years away. What they actually saw was a rhythmic shifting of the star's spectral lines. Like all stars, 51 Peg has gases in its atmosphere that intercept specific wavelengths of light and keep them from reaching Earth. When you smear the starlight into its constituent colors with a spectrometer, those colors intercepted by the star's gases are absent and instead appear as black lines in an otherwise rainbowlike spectrum. If a star is moving, those lines shift in position. If the star is moving toward us, the Doppler effect will shorten its light waves, shifting the black lines toward the blue end of the spectrum. Similarly, if the star is moving away from us, the light waves will lengthen and the black lines shift toward the red. And it's just such shifts—first to the blue, then to the red, over and over—that the astronomers saw. Their conclusion: An orbiting planet is gently tugging the star to and fro.

Gray's analysis of 51 Peg's light, though, shows that the "motion" is actually a hitherto unsuspected pulsation in the star itself, which is skewing the lines. Despite Gray's self-assurance, however, his argument is hardly ironclad. "David Gray is an extremely careful observer," says Sallie Baliunas of Harvard, an expert on, among other things, stellar pulsations. "You have to take what he says seriously. But while his interpretation is not impossible, it's a long way from being convincing."

In any case, Gray's observations don't apply to any of the planets found after 51 Peg (there are anywhere from 8 to 11 more, depending on whom you ask, meaning that there may now be more planets known outside the solar system than in). Most astronomers still think, in short, that the planets nobody ordered still need explaining.

Although they'd been scooped by Mayor and Queloz, San Francisco State University astronomers Geoff Marcy and Paul Butler made up for lost time. In the 12 months following the 51 Peg announcement, the pair found no fewer than six more planets in the northern skies. Most of them, like 51 Peg, had not been on anyone's menu. Three—around 55 Rho 1 Cancri, 44 light-years away in the constellation Cancer; Tau Bootes, 49 light-years away; and Upsilon Andromedae, 54 light-years away—were very much like that first peculiar discovery: absurdly close in for their Jupiteresque mass. But two others—around 70 Virginis, 59 light-years away, and HD 114762, 91 light-years away, both in Virgo—were equally strange in another way. While they orbited at a slightly more conventional distance from their respective stars, their orbits were highly eccentric—not the nearly circular orbits of Jupiter and Saturn but more elongated.

One possible explanation would be that these are not planets at all but brown dwarf stars. A brown dwarf is about as close as a star can get to being a planet without actually being one. For one thing, it is small enough to be mistaken for a very large planet, at least from a distance. And since, being small, its core is not put under enough pressure for nuclear fusion reactions to take hold, it emits only a dim glow at first and then gradually goes dark. Despite its planetlike appearance, however, there's no escaping its lineage: a brown dwarf is formed directly from a collapsing gas cloud—a stellar process if ever there was one—rather than from the accretion of dust and gas that gives birth to planets. Theorists have always assumed that brown dwarfs must be at least ten times more massive than Jupiter and probably more. But perhaps they've overestimated. Maybe a brown dwarf can be six times as massive as Jupiter or even less. If so, it could be that 70 Virginis and HD 114762 are simply double-star systems, each with one real star and one almost-star. This would explain the eccentric orbits, at least: double stars usually orbit each other in distinctly noncircular paths.

That, however, doesn't explain 16 Cygni B. It is the most eccentric of the bunch but has only one and a half times Jupiter's mass. Even though theorists are prepared to countenance smaller-than-expected brown dwarfs, there are some limits. "It's really stretching things to call that a brown dwarf," says Fred Rasio, a theorist at MIT. Time for another theory.

As it happens, Rasio has one. "We start with the idea that it's plausible to form three or four or five Jupiters in a young solar system," he says; that's pretty much inevitable if

One possible explanation would be that these are not planets at all but brown dwarf stars. A brown dwarf is about as close as a star can get to being a planet without actually being one.

you've got enough raw material. With that many giant planets around, though, they'll certainly be getting in each other's way. Based on computer models, there are two things that can happen. About half the time, two or more Jupiters will collide and fuse into a single planet, and the collision will skew the original planets' circular orbits into an elliptical one. Presto: 70 Virginis, HD 114762, and 16 Cygni B (which, truth be told, would have to have been made from two Saturns rather than two Jupiters, but that's reasonable).

The other half of the time, the planets will merely have a close encounter. When that happens, one of the planets will be flung out of the system entirely and sent wandering throughout the galaxy. (Such a rogue Jupiter would be difficult to observe, since it would be small and dim, unless it happened to invade our solar system, a prospect too distressing and unlikely to contemplate.) The other big planet would be slung in toward the star, taking up a highly eccentric, cometlike orbit. Such an orbit cannot last long. Each time the planet whips in close to the star, its own gravity is great enough to distort the parent star's shape very slightly, which robs the planet of a little bit of energy. It's like an out-of-control skateboarder grabbing at a signpost to slow down. After millions of years, the planet ends up in a circular orbit close in to the star. Voila: 51 Peg and its brethren.

If he's right, Rasio has not only explained just about all of the new planet observations in one shot but may have shown that systems like ours are the exception, not the rule. Of all the new-planet discoveries, only one—around 47 Ursae Majoris—has a large planet in a roughly circular orbit at a somewhat Jupiterlike distance. "It's really much too early to generalize," Rasio says. "But at some level, you get the impression that most solar systems are not like ours."

They are, moreover, unlike ours in a most inhospitable way. Big planets in eccentric orbits will sooner or later disrupt the orbits of small, friendly places like Earth; the greater the eccentricity, the more quickly the planet will swoop in and disrupt things. Even big planets that spiral slowly inward in nearly circular orbits, as in Lin's scenario, will fling Earth-like planets out as they go. Either way, you wouldn't expect life to arise and survive, and so we may be more alone in the universe than we like to think.

But wait! Doug Lin has already cooked up a happier scheme. Yes, big planets migrate in, and they fling Earths out into deep space. But what's to stop new Earths from forming? In fact, Lin thinks the whole thing may happen more than once. There may have been four, five, six Jupiters in earlier versions of our own solar system, each swallowed by the sun. This Jupiter is only the latest. If so, then Earths are plentiful after all. Life is abundant. All is well.

It is until the next observation comes in, that is. Brilliant as these theorists are at explaining each new unexpected fact, they're still working with very few data. That's better than zero, of course, but even the theorists admit they could use much more information. "Right now," says theorist Jack Lissauer, of the NASA Ames Research Center, "we're seeing individual planets. But our understanding of how planets form in this solar system doesn't come just from Earth. It's from looking at all the planets."

The theorists may not have to wait long. Even now, Marcy and Butler have moved their planet-hunting operation from small telescopes at the Lick Observatory near San Jose, California, to the powerful Keck telescope in Hawaii, where they'll be able to find many more, and smaller, planets. Other planet hunters are at the Keck as well. Of course, planets at Jupiterlike distances take a dozen years or so to complete a single orbit, which means you'd have to observe them that long to be sure they are really there. So it could take decades before we know for certain which theory of planetary formation is most plausible, or how rare a particular sort of system is. Until then, though, it's still fun to watch the theorists at play.

JPL's Interferometry Mission To Hunt for New Planets[7]

The Space Interferometry Mission (SIM) is a stepping-stone in NASA's Origins program that is investigating the 15 billion-year-long chain of events that began with the Big Bang and produced the Universe.

The Jet Propulsion Laboratory is expected to award a contract about September 1 for a space mission to hunt for other planets in the Milky Way. The flight will require such precise measurements that its principal instrument will be calibrated to the subatomic level.

The Space Interferometry Mission (SIM) is a stepping-stone in NASA's Origins program that is investigating the 15 billion-year-long chain of events that began with the Big Bang and produced the Universe.

SIM will be the first long baseline optical interferometry instrument in space. Funding for development of the interferometer instrumentation and the spacecraft is capped at $480 million. Launch is set for 2005 on an Evolved Expendable Launch Vehicle, and SIM is expected to operate from the L-2 Lagrangian point.

In a seven-year mission, it will look at the 100 or so nearest stars in the galaxy, pinpointing their location to an accuracy of 4 microarcsec., which is 250 times more precise than the best available star catalog. It will determine the angular positions and intrinsic luminosities of stars in the Milky Way and, using rotational parallax, the distance to other galaxies. And it will help scientists understand the dynamics of globular clusters and spiral arms in galaxies.

But its central goal is to improve on current methods of detecting possible planets in orbit around stars like our Sun. To do so, it will use the common practice of looking at the "wobble" such planets induce in the star through their gravitational pull. Planet candidates have been spotted previously. What SIM should provide is a refinement of the possible planets seen. Candidates thus far are large, gaseous planets like Jupiter, SIM Program Director Farooz Naderi of JPL explained. Current instruments are not refined enough to detect the subtle gravitation shifts that smaller, Earth-sized planets would cause. Since these planets are the most likely to harbor life, they are of especially high interest.

The technical challenge is to detect the small planets. While SIM will not be looking across the universe, as the Hubble Space Telescope does, the Milky Way Galaxy is still 100 light-years across.

---

7. Article by Michael Mecham from *Aviation Week and Space Technology* p54 + Aug. 24, 1998. Copyright 1998 *Aviation Week and Space Technology*. Reprinted with permission.

The reference design for the spacecraft calls for it to combine light waves from two sets of 30-cm. (1-ft.) diameter telescopes (a fourth mirror is included in each instrument cluster as a backup). The telescopes are clustered at opposite ends of a 10-meter (33-ft.) boom. If the interferometer can be pointed with sufficient accuracy it will have the effective resolution of a l0-meter telescope. It should be able to pinpoint stellar objects to about 2-3 orders of magnitude better than the state-of-the-art *Hipparcos* spacecraft, launched by the European Space Agency in 1989, Naderi said.

Since the detector is measuring the time travel of light, the exact distance between it and the two "stereo" sets of cluster telescopes on the ends of the boom must be known. Space-based radio interferometers use the same technique, but radio waves are relatively long. Optical wave lengths are subatomic. Furthermore, the distance between the detector and telescopes will be dynamic since the spacecraft will be subject to expansion and contraction heating cycles in space.

SIM's tolerance level will be 50 picometers. "To give you an idea of how small that is, a hydrogen atom is 100 picometers," Naderi said. The space imaging processing will be done partly on board by the detector and partly at JPL.

Design of the instrument is not a done deal. "The SIM program has some daunting mechanical challenges," said Chuck Rudiger, who heads Lockheed Martin Missiles & Space's advanced space observatories business development program.

Exactly who is competing to build SIM and its interferometer is a partial mystery, at least officially. Naderi said he is prohibited from revealing who has bid, even though the request for proposals (RFP) deadline has passed. The likely candidates are LMMS, TRW and Ball Aerospace, since all three participated in the Pre-Phase A study to define the SIM reference design.

Lockheed Martin and TRW have acknowledged that they are competing, but an official for Ball refused comment.

The competition is to design the spacecraft, to join a JPL-led team to design the SIM instrument package, or to do both. The dollar value of the spacecraft and instrument are roughly equal, Rudiger said.

Lockheed Martin is seeking both contracts; TRW cites competitive factors for not saying which it is bidding on. One likely scenario is that an industry team will be chosen, according to Greg Davidson, director of TRW's space science programs.

"Space interferometry is critical for the future of astronomy and has a very high potential for a variety of future applications," Davidson said. "There are a number of interferometry missions now and in the future that we want to be a part of."

Deep Space-3, one of NASA's New Millennium series of small spacecraft, is another prime candidate at the moment. A precursor to SIM, it is to be launched in about three years to demonstrate optical interferometry in space. Instead of separating telescopes on a long boom as SIM does, *DS-3* will put them on two separate satellites that will be flown in formation. A third spacecraft will carry the detector instrument package. Lockheed Martin and TRW are among those awaiting the RFP for *DS-3*, which is expected as early as September.

More advanced Origins spacecraft are still to come. They include the *Next Generation Space Telescope*, set for launch in 2007. Not a replacement for *Hubble*, the *NGST* will operate in the infrared and achieve a higher resolving power by taking advantage of advances in lightweight optics and electronics that have occurred since *HST* was designed in the 1970s.

Planet Finder, a 2011 mission, is to combine technology from *NGST* and SIM to actually image planets. And scientists hope that a successor to *Planet Finder* called *Planet Imager* will deliver Landsat-quality images of such planets. Its launch would be about 2018.

Bibliography

Books and Pamphlets

Bingham, Caroline. *Incredible Universe.* SnapShot, 1995.

Burgess, Eric. *Outpost on Apollo's Moon.* New York: Columbia University Press, 1993.

Burrows, William E. *This New Ocean: The Story of the First Space Age.* New York: Random House, 1998.

Clay, Rebecca. *Space Travel and Exploration.* New York: 21st Century Books, 1997.

Cozic, Charles P. *Space Exploration: Opposing Viewpoints.* Greenhaven Press, 1992.

Curtis, Anthony R. *Space Almanac.* Gulf, 1992.

Gribbin, John R. and Goodwin, Simon. *Empire of the Sun: Planets and Moons of the Solar System.* New York: New York University Press, 1998.

Hardersen, Paul S. *The Case for Space: Who Benefits from Explorations of the Last Frontier?* ATL Press, 1997.

Johnson-Freese, Joan and Handberg, Roger. *Space, The Dormant Frontier: Changing the Paradigm for the 21st Century.* Praeger Pubs., 1997.

Launius, Roger D. *Frontiers of Space Exploration.* Greenwood Press, 1998.

Levine, Alan J. *The Missile and Space Race.* Praeger Pubs., 1994.

Lewis, John S. *Mining the Sky: Untold Riches from the Asteroids, Comets, and Planets.* New York: Addison-Wesley, 1996.

Madders, Kevin. *A New Force at a New Frontier: Europe's Development in the Space Field in the Light of Its Main Actors, Policies, Law, and Activities from Its Beginnings Up to the Present.* Cambridge University Press, 1997.

Moore, Patrick. *Space Travel for the Beginner.* New York: Cambridge University Press, 1992.

Neal, Valerie. *Where Next, Columbus?: The Future of Space Exploration.* Oxford University Press, 1994.

Reeves, Robert. *The Superpower Space Race: An Explosive Rivalry Through the Solar System.* Plenum Press, 1994.

Sagan, Carl. *Pale Blue Dot: A Vision of the Human Future in Space.* Headline Book Pub., 1995.

Schmidt, Stanley, and Zubrin, Robert. *Islands in the Sky: Bold New Ideas for Colonizing Space.* Wiley, 1996.

Stott, Carole and Gorton, Steve. *Space Exploration.* New York: Knopf, 1997.

Tesar, Jenny E. *Space Travel.* Heinemann Interactive Library, 1998.

Twist, Clint. *Gagarin and Armstrong.* Raintree Steck Vaughn Pubs., 1995.

Verba, Joan-Marie. *Voyager: Exploring the Outer Planets.* Lerner Publs., 1991.

White, Frank. *The Overview Effect: Space Exploration and Human Evolution.* American Institute of Aeronautics and Astronautics, 1998.

Additional Periodical Articles with Abstracts

For those who wish to read more widely on the subject of Space Exploration, this section contains abstracts of additional articles that bear on the topic. Readers who require a comprehensive list of materials are advised to consult *Reader's Guide Abstracts* and other Wilson indexes.

On John Glenn:

John Glenn's Excellent Adventure. Mark Alpert. *Scientific American* v. 280 no. 1 p. 30+ Jan. '99.

Senator John Glenn's return to space, at the age of 77, may have been a boost to NASA's media profile, but it also had scientific merit. The Discovery STS-95 mission gained significant publicity because millions of TV viewers avidly followed the progress of the man who had been the first American to orbit Earth. Medical experiments conducted during the mission should help examine the fascinating parallels between aging and space flight.

The Astronaut, Down to Earth. Matt Bai. *Newsweek* v. 132 no. 23 p.58 Dec. 7 '98

An interview with astronaut and former senator John Glenn. Among the topics discussed are how he has adjusted since his return to Earth, whether he ever got sick in space, and whether looking down on Earth changed his political views.

Sky High. *People Weekly* v. 50 no. 20 p.64-5 Nov. 30 '98

John Glenn, the oldest man ever to go into space, returned home recently. Glenn, who was the first American to orbit Earth in 1962, went into space again for nine days aboard the space shuttle *Discovery* at the age of 77. The four-term senator and ex-fighter pilot was clearly in awe at the reception he got as an estimated 500,000 turned out to see him parade through lower Manhattan on his victory lap. After the New York welcome, Glenn and his wife left for Houston, Texas, where doctors will test him in order to assess the effects of the flight.

As Glenn Suits Up for Launch Again, Is America Focusing on the Right Stuff? Chuck McCutcheon. *CQ Weekly* v. 56 no. 42 p.2891-92 Oct. 24 '98

Seventy-seven-year-old Senator John Glenn, D-Ohio, will board the shuttle *Discovery* convinced that the focus of attention for his nine-day flight should be on the science of how space travel influences the aging process, not on his image as an authentic American legend. October 29 is the scheduled date for the launch. As much as he has attempted to prevent it, Glenn's trip has snowballed into a cultural phenomenon because of who is he and what he has already accomplished, not because of any experiments he will be doing. The spectacle of the first American to orbit the Earth back in 1962 to become the oldest person in space has captured the attention of a public that has become blasé about shuttle flights.

Still the Right Stuff. Al Reinert. *TV Guide* v. 46 no. 43 p.14-18+ Oct. 24-30 '98

Thirty-six years after America's first manned trip into space, pioneer astronaut John Glenn and former news anchor Walter Cronkite will return to the roles that made them legends. The space shuttle *Discovery* will depart the Kennedy Space Center at Cape Canaveral, Florida, on October 29, and Senator John Glenn will be part of the crew, going into orbit for the second time in his 77 years, by far the oldest person ever in space. To celebrate the event, Cronkite will return to the nation's television screens, steering viewers through space as he coanchors CNN's live coverage of the 95th shuttle mission. From his desk in the CBS news broadcast booth at Cape Canaveral, Cronkite told the story of the 1962 adventure more cogently than the space hero himself. That historic mission and both men's roles in it are discussed.

John Glenn Flies Us Back to the Age of Innocence. Richard B. Stolley. *Life* v. 21 no. 11 p.50-5 Oct. '98

At the age of 77, John Glenn, who in 1962 became the first American to orbit the Earth as part of Project Mercury, is returning to space for nine days starting on October 29. After his first space flight, NASA officials grounded him because they were convinced that he was worth more selling space travel than undertaking it. After 18 months, however, he became discouraged, quit NASA, and ended up in politics, becoming a four-term Democratic senator. Around three years ago, as the growing number of elderly Americans started to attract scientific attention, Glenn sensed a chance to get back into space. He recruited the help of the National Institute on Aging, which supported the notion that knowledge gained from studying a senior citizen in space would be valuable. Glenn's preparations for the space flight are discussed.

Can John Glenn Do It Again? Helen Thrope. *Texas Monthly* v. 26 no. 10 p.104-7 + Oct. '98

John Glenn once saved America from losing the space race to the Russians, but now he has the task of saving NASA itself. Although the official purpose of the 77-year-old's shuttle mission is to examine parallels between the aging process and the effects of space travel on the human body, the reality is that NASA longs for a return to its glory days, and there is no better way to generate public support for, say, a manned mission to Mars, than to remind the American people of the first time the nation fell head over heels for an astronaut. The writer discusses Glenn's career and the agenda of NASA administrator Dan Goldin.

The Ripe Stuff. Michael O'Neill; Rinker Buck. *Vanity Fair* no. 458 p.282-93 Oct. '98

As a once-fabled NASA turns 40, 77-year-old John Glenn's October return to space may reignite America's passion for the final frontier. The astronaut-senator, who made history in 1962 in a Mercury capsule named *Friendship 7*, will return to space for nine days aboard the space shuttle. Space officials decided in January that the event would position NASA for the 1990s, a decade when Americans appear to evaluate themselves by the intensity of their nostalgia rather than their astounding technical achievements. Despite rumors that space scientists doubt the practical value of sending a man of his age into space—Glenn will conduct geriatric-related experiments—America appears more than willing to accept this victory of symbolism over substance. Photographs of a number of individuals involved in some way with the Mercury-Gemini era are presented.

On the Exploration of Mars:

Mars Lander Liftoff Advances Bold New Exploration Strategy. Craig Covault. *Aviation Week and Space Technology* v. 150 no. 2 p.434-5 Jan. 11 '99

The U.S. Mars Polar Lander and its twin Deep Space 2 penetrators are on their way to Mars on a mission that will help usher in a new international Mars strategy second in scope solely to worldwide aerospace industry collaboration on the International Space Station. The aim of the mission is to unite the U.S., France, Italy, the European Space Agency, and Russia on an unmanned Mars surface-exploration program that will result in several hundred million dollars' worth of new space technology and deal directly with the issue of whether there was, or still is, life on Mars. It is planned that the new strategy will be composed of up to 12 multinational launches—four of which have already been completed—through 2009.

Donna Shirley. *Current Biography* v. 59 no. 8 p.51-5 Aug. '98

Donna Shirley is the director of the Mars Exploration Program at NASA's Jet Propulsion Laboratory. In 1992, she became the manager of the team that went on to design Sojourner, the robotic rover that accompanied the spacecraft *Pathfinder* on its flight to Mars in 1997. When the craft landed on Mars in July, Sojourner became the first autonomous robot to explore the surface of a planet other than Earth. It was Shirley who came up with the basic design concept for the rover, as well as the idea of sending it to Mars aboard *Pathfinder*. She thus experienced the mission not only as a victory for space exploration but also as a personal triumph.

Mirror-Machine to Mars? Terry Kammash. *Ad Astra* v. 10 no. 2 p.34-7 Mar./Apr. '98

The tremendously promising Gasdynamic Mirror (GDM) Fusion Propulsion System employs a magnetic "bottle" to keep fuel in an ionized form for a period long enough to permit nuclear fission reactions. Craft powered by fusion power plants like GDM are the ultimate embodiment of quicker, better, and cheaper space exploration. Moreover, really practical exploration of the solar system and colonization on a major scale and at an affordable expense will only be possible with vehicles of this kind. The writer explains how the system works.

Technology Initiative to Advance Exploration. Craig Covault. *Aviation Week and Space Technology* v. 148 no. 18 p.40-2 May 4 '98

NASA's X2000 Advanced Deep Space System Development Program aims to reinvigorate the development of the space technology required for five new deep-space missions. The program is being run by the Jet Propulsion Laboratory. It specifically aims to support development of the Solar Probe, the Deep Space 4 mission, the Mars sample return, the Pluto flyby, and the Europa orbiter.

French Involvement May Boost Mars Studies. Ron Cowen. *Science News* v. 153 no. 13 p.197 Mar. 28 '98

The French space agency and NASA are proposing to collaborate on missions to drill into Mars's northern lowlands, which recent evidence indicates may have been formed by an ancient ocean. In a plan currently being negotiated, the French would spend some $400 million on Mars exploration, almost doubling the U.S. budget for recovering samples over the next decade and providing several new, small missions for studying the Red Planet. The French space agency would also finance a number of launches of the recently developed Ariane-5 rocket and supply most of the parts for a Mars orbiter, which is due to transport Martian soil and rock samples back to Earth in 2008. In return, NASA would donate some of the samples to French researchers.

Goldin Reverses Cuts to Manned Moon/Mars Plan. Joseph C. Anselmo; Craig Covault. *Aviation Week and Space Technology* v. 148 p.24-5 Feb. 2 '98

NASA administrator Daniel S. Goldin has reversed a controversial command to cancel the majority of agency research programs targeted at supporting further astronaut missions to the Moon and Mars early in the next century. Goldin's directive nullifies much of a memo to NASA field centers on January 9 ordering the termination by January 30 of projects uniquely aimed toward human exploration beyond low-Earth orbit. The memo from Richard J. Wisniewski, the deputy associate administrator for space flight, stated the terminations, which involved around $10 million worth of activities, were being directed to continue to the settlement of funding deficits within the agency. Cost overruns on the International Space Station program have left NASA frantically trying to find savings in its fiscal 1998 budget, which runs through September 30.

On the International Space Station:

Flywheels Show Promise for 'High-Pulse' Satellites. Paul Proctor. *Aviation Week and Space Technology* v. 150 no4 p. 67 Jan. 25 '99

Advanced flywheel technology is being prepared for tests on the International Space Station that, if successful, could significantly further future space vehicle performance and payload capability and be mission-enabling for others. Potential future flywheel applications include "high-pulse" low- and medium-Earth orbit communications satellite constellations as well as space-based lasers and radars. Flywheels optimized for space applications could provide longer life, a better than 2:1 weight advantage, and more compact energy storage than any practical chemical battery-based system, says John R. Barron, program manager for the Attitude Control and Energy storage Experiment.

1998: A Space Odyssey. Mark Nichols. *Maclean's* v. 111 no. 48 p.81 Nov. 30 '98

Construction of the International Space Station is under way. The huge project, scheduled for completion in 2004, is a 16-nation undertaking led by the United States, with Russia as a major partner and 11 European nations, Brazil, Canada, and Japan contributing equipment and funds. The 450-ton station will be a network of laboratories, living space, service areas, and immense solar panels. A Russian Proton rocket named Zarya was recently launched from Baikonur in southern Kazakhstan, and from now on, 45 U.S. and Russian missions will add parts to it to build the 450-tonne station. The Canadian contributions to the station are discussed.

Putting It Together. Adam Rogers. *Newsweek* v. 132 no. 25 p.74-5 Dec. 21 '98

Five U.S. astronauts and a Russian cosmonaut have successfully connected the initial two pieces of the International Space Station. During three spacewalks, they went on to turn on the lights, computers, and air conditioning in these sections. Photographs taken during the assembly process are presented.

Europe Advances ATV Development for ISS. Michael A. Taverna. *Aviation Week and Space Technology* v. 149 no. 22 p.29 Nov. 30 '98

The European Space Agency has finalized agreements for full-scale development of the Automated Transfer Vehicle (ATV). A critical component of the International Space Station, the ATV, in conjunction with the Ariane 5 heavy-lift booster, will resupply the complex in cargo and fuel and periodically reboost it to proper orbit. The unit will have its own propulsion system and be designed for up to eight months in orbit, but it will not be recoverable. The recently signed agreements covered a $470 million contract with Aerospatiale, ATV development prime; a $23 million framework agreement with the Russian Space Agency and RSC Energia to integrate the ATV system into the ISS; and a $30 million contract with French Space Agency CNES to develop interfaces with the Ariane 5 Evolution heavy booster.

Who Needs This? Jeffrey Kluger. *Time* v. 152 no. 21 p.88-91 Nov. 23 '98

The International Space Station (ISS) might be a massive waste of money. On November 20, the first part of the 16-country, NASA-led project is due to be launched from the Baikonur space center in Kazakhstan. The station is between 3 and 12 times over budget and is being kept going mainly by American funds. In addition, the project is running 14 years late and will probably fall further behind before work is completed. Worst of all, there are serious concerns over whether there will be anything genuinely useful for the ISS to do once it has gone into orbit. Nonetheless, the ISS project seems likely to continue and might dominate NASA's exploratory agenda for a generation.

NASA's Mission to Nowhere. Timothy Ferris. *New York Times* (Late New York Edition) p. 9 Sec 4 Nov. 29 '98

NASA's planned International Space Station is a step in the wrong direction. It is bad news for NASA, science, and for the manned space effort it was once expected to advance. The project will eat up money but be of almost no use in getting to Mars, the Moon, or elsewhere.

On Private Enterprise and Space Exploration:

Water on the Moon. *Discover* v. 19 no12 p. 22 Dec. '98

Estimates for the amount of water on the Moon have increased recently. In March, *Lunar Prospector* detected evidence of as much as 300 million tons of water at the Moon's poles. Further data and analysis suggest, however, that Earth's satellite may hold over 600 billion tons of water, which is 2,000 times more than the original estimate. Space physicist David Lawrence of Los Alamos National Laboratory, who helped analyze the data, says most of the water lies 16 inches below the lunar surface, and he suspects that much of it is in solid chunks of ice rather than the dusty frost suggested by earlier reports.

The Moon Rediscovered. Jeff Foust. *Sky and Telescope* v. 96 no6 p. 32-4 Dec. '98

Recent data that has returned from NASA's low-budget Discovery-series Lunar Prospector spacecraft demonstrates how much more there is to still to learn about the Moon. The findings, published in the September 4th issue of *Science*, have afforded new insights and raised new questions about the nature and origin of the

Moon. The writer discusses the significance of the returned data.

Launching for Dollars. Beth Elliott. *Ad Astra* v. 10 no4 p. 27-9 July/Aug. '98

Applied Space Resources (ASR) is a company that is planning to launch the first commercial mission for the return of lunar samples. The company was established with the objective of employing existing technologies to deliver spacecraft to any destination with accuracy and bring back resources and data with equal accuracy, but for a profit. ASR's initial goal is to launch a craft that will gather and return a quantity of the loose material to be found on the surface of the Moon, which is known as lunar regolith, to be sold to governments, research groups, and even ordinary members of the public. The craft, the *Lunar Retriever*, will carry a main payload of a sample-return capsule and robotic devices to load the capsule. It will deploy video equipment that will transmit to Earth live pictures of the *Lunar Retriever*'s activities on the surface and its departure for the return trip.

Mining Water on the Moon. Jim Wilson. *Popular Mechanics* v. 175 no7 p. 32-3 July '98

The recent discovery by NASA's *Lunar Prospector* of water on the Moon has increased excitement about putting the resource to work. In theory, the water can be easily turned into air to breathe, hydrogen for fuel, and oxygen to burn it, and NASA estimates that there is enough lunar water to support over 1 million space shuttle launches. Encouraged by this, astronomers, planetary scientists, and deep-space explorers are contemplating lunar-based stations that would enhance their ability to explore space. The writer examines a proposal from space architect Madhu Thangavelu of the University of Southern California Los Angeles for a space rover, the Nomad Explorer, which would make lunar exploration easier than before.

Company Targets Asteroid—And Profits. Andrew Lawler. *Science* v. 277 p. 1756 Sept. 19 '97

SpaceDev, a Steamboat Springs, Colorado-based company, is planning to launch the first privately financed mission beyond Earth's orbit. The company intends to send a probe, called Near Earth Asteroid Prospector (NEAP), to one of thousands of small asteroids and comets that pass close to Earth's orbit and to sell the data collected about the asteroid's composition to customers, including NASA. According to SpaceDev, the NEAP probe, which would weigh about 300 kilograms and carry a number of well-tested instruments, can be built, launched, and operated for less than $50 million, a fraction of what it would cost NASA.

Company Plans Asteroid Visit. Joseph C. Anselmo. *Aviation Week and Space Technology* v. 147 p. 94-5 Sept. 15 '97

SpaceDev, a new commercial venture, has announced plans to land a small, privately funded spacecraft on a near-Earth asteroid in 2000 to gather scientific information and stake a claim for future mining rights. The Steamboat Springs, Colorado, company hopes the sale of scientific data will finance the mission, which will likely cost less than $50 million. It is the first time a company has not sought government funding for such a venture. SpaceDev founder James W. Benson says the design for the spacecraft, dubbed the Near Earth Asteroid Prospector, should be ready by the end of 1997.

On New Technologies and Discoveries:

Shading the Twinkle. Gary Stix. *Scientific American* v. 279 no6 p. 40 Dec. '98

Researchers have taken a major step toward producing an instrument capable of photographing new planets. Astronomers have inferred the existence of a dozen or so planets outside the solar system from the wobble in light observed by telescopes as a planet orbits around a nearby star and exerts its gravitational pull on the gaseous body. A parent star, millions of times brighter than a planet, can drown the lesser image, however. In a recent issue of Nature, Philip M. Hinz and his colleagues at the Center for Astronomical Adaptive Optics at the University of Arizona, Tucson, describe a starlight shading device, called a nulling interferometer, that is fitted to the Multiple Mirror Telescope on Mount Hopkins in Arizona. The device, which consists of two mirrors mounted five meters apart on a rigid frame, cancels out the light from a star when the star is at an exact right angle to the frame.

Long Firing of Ion Engine Meets DS1 Requirements. Michael A. Dornheim. *Aviation Week and Space Technology* v. 149 no. 23 p.82 Dec. 7 '98

The test run of Deep Space 1's xenon ion rocket has been very "clean," said John F. Stocky, the Jet Propulsion Laboratory manager of the NSTAR solar electric propulsion project. The rocket is well into a 13-day continuous firing, which is set to end on December 7, and has passed the 200-hour benchmark necessary to declare the technology validation mission a success.

Callisto's Icy Secret. Charles Petit. *U.S. News and World Report* v. 125 no. 17 p.55 Nov. 2 '98

Researchers at the University of California-Los Angeles believe that data provided by NASA's Galileo spacecraft suggests that Jupiter's moon Callisto may have an ocean locked beneath its frozen surface. NASA's Jet Propulsion Laboratory in Pasadena, California, is already planning to send more spacecraft to the Jovian moon Europa, where oceans already seem likely, to check in greater detail for signs of life. Scientists say that the chances of Callisto supporting life are slimmer, given that its ocean probably lies beneath about 100 miles of solid ice. Nonetheless, the prevalence of oceans orbiting the Sun suggests that Earth is not unique and that many similar worlds accompany other stars.

Flying with Ion Power. Leon Jaroff. *Time* v. 152 no. 15 p.52 Oct. 12 '98

With Deep Space 1, NASA will launch an intelligent new generation of spacecraft. Although it resembles many other unmanned probes that NASA has launched, Deep Space 1 will be guided by an electronic brain and powered by an ion propulsion engine. Flight planners hope that the craft will make a number of interesting discoveries about the target asteroid, 1992KD, such as its composition and the structure of its surface. The ship's main task, however, will be to validate a wide range of new technologies that NASA had always regarded as too dangerous to be used on a high-profile mission.

A Dozen New Planets . . . And Still Counting. Ron Cowen. *Science News* v. 154 no. 13 p.197 Sept. 26 '98

The discovery of two new planets orbiting stars similar to the Sun was announced at the Carnegie Institution of Washington, D.C., on September 9 by R. Paul Butler of the Anglo-Australian Observatory in Epping, Australia. A planet orbiting the star HD210277, discovered by the Keck 1 Telescope on Hawaii's Mauna Kea, is the first planet whose average distance from its parent star is almost the same as Earth's distance from the Sun. The other new planet, orbiting the star HD187123, is nearer to its host than any other planet found yet. Butler and his colleagues will report details in Publications of the Astronomical Society of the Pacific.

Eking Out a Life in the Ice. Adam Rogers. *Newsweek* v. 132 no1 p. 62 July 6 '98

The survival of microbes in Antarctica lends support to the theory that organisms can survive anywhere as long as liquid water is present. John Priscu, a microbial ecologist at Montana State University, and colleagues have described an aggregate community of microbes living in oases of liquid water inside a frozen lake in Antarctica. Exobiologists believe that if life can make it there, it could make it in places such as Mars or the ice-covered seas of Jupiter's moon Europa.

Hello! Is There Anybody Out There? Judith Braffman-Miller. *USA Today* v. 126 no2632 p. 30-3 Jan. '98

It is a fascinating possibility that there are other small, watery worlds that resemble Earth orbiting far-distant stars. At a January 1996 meeting of the American Astronomical Society, Geoffrey W. Marcy of San Francisco State University and R. Paul Butler of the University of California, Berkeley, reported that they had spotted two new planets outside Earth's solar system. One, nicknamed "Goldilocks," orbits at just the right distance from its parent star for liquid water to exist on the surface. The other may have liquid water, but only in its atmosphere. Water is important because it is believed to have contributed to the formation of life on Earth. The writer discusses the discoveries of other extrasolar planets.

Shuttling into the 21st Century. Donald F. Robertson. *Astronomy* v. 23 p. 32-9 Aug. '95

In cooperation with private aerospace companies, NASA is making plans for a new, reusable, single-stage-to-orbit (SSTO) launch vehicle to replace the aging space shuttle. Because much of the space agency's budget is tied up in shuttle operations, it has little cash for development of a second-generation shuttle that might cost less to operate. NASA Administrator Daniel Goldin, however, has found a few hundred million dollars in the budget for preliminary development of a new vehicle, provided industry contributes matching funds. In March, the agency selected three firms to draw up proposals for a test vehicle; the White House and Congress will decide by early summer 1996 whether to proceed further. Possible designs for new launch vehicles are described, and a sidebar discusses ways that the current shuttle fleet could be upgraded and kept flying, possibly through the establishment of a government-chartered corporation.

Index

Delaney, John 126, 129
Delbaere, Louis 14
diabetes
    experiments in space 14
Discovery (space shuttle)
    John Glenn and 3–10, 11–15, 16–
        18, 22–24
Duitch, Ellis 12

Easterbrook, Gregg
    "Cosmic Clunker" 80–82
Eclipse Astroliner 120, 121
Eclipse Sprint 121
Electron Reflectrometer (ER) 43
Europa (moon) 126–129
    Galileo space probe 131
Evolved Expendable Launch Vehicle
140

Farquhar, Bob 98
Fender, Donna 51
Field Robotics Center 100
Flamsteed, Sam
    "Impossible Planets" 133–139

Galileo mission 129
Galileo space probe 131
Garneau, Marc 12
Gideonse, Theodore
    "Looking For Money On the Lunar
        Surface" 102–103
Gilruth, Bob 5
Glenn, John
    space shuttle *Discovery* flight 3–10,
        11–15, 16–18, 22–24
Gold, Thomas 127, 128
Goldin, Daniel 4, 6, 28, 49, 60, 80, 81,
96
    NEAP and 87, 91, 92
Golombek, Matthew 33, 37, 38
Gray, David 136, 137
Gruen, Adam 59
Gump, David 100

Hadfield, Chris 12
Harper, Lynn 89
Hartsfield, James 15
Harvest Moon 100
Hawaleshka, Danylo
    "A Hero in Orbit" 11–15
HD 114762 137, 138
Helfand, David 77
helium
    mining, on lunar surface 103
Hipparcos spacecraft 141
Holland, Al 72
Hope, Dennis 103
Hubble Space Telescope 8, 140
    repair mission 64
Huntress, Wesley 28, 91, 113

Imager for Mars Pathfinder 33
International Space Station 12, 14, 17,
51, 74–79, 80–82
    assembling 64
    assembly schedule 62
    cost of 75, 80
    Functional Cargo Block(FCB) 70
    history 57–67
    Node 1 71
    public support 79
    Service Module 71
    U.S. Laboratory Module 72
    value of 83–84
ion propulsion engine
    Deep Space 1 113–116

Johnson, Torrence 129
Joosten, Kent 102
Jupiter
    Europa (moon) 126–129, 131
    Io (moon) 128

Kelly, Michael 121
Kennedy, John F.
    John Glenn and 5
Khedouri, Frederick N. 76